OPTIMIZING HVAC SYSTEMS

OPTIMIZING HVAC SYSTEMS

Albert Thumann, P.E., CEM

Published by
THE FAIRMONT PRESS, INC.
700 Indian Trail
Lilburn, GA 30247

Library of Congress Cataloging-in-Publication Data

Optimizing HVAC systems.

   Includes index.
   1. Heating—Equipment and supplies. 2. Ventilation—
Equipment and supplies. 3. Air conditioning—Equipment
and supplies. I. Thumann, Albert.
TH7345.068   1987   697   87-45326
ISBN 0-88173-042-4

Optimizing HVAC Systems
©1988 by The Fairmont Press, Inc. All rights reserved. No part of this publication may be reproduced or transmitted in any form or by any means, electronic or mechanical, including photocopy, recording, or any information storage and retrieval system, without permission in writing from the publisher.

Published by The Fairmont Press, Inc.
700 Indian Trail
Lilburn, GA 30247

ISBN 0-88173-042-4   FP

ISBN 0-13-638305-X   PH

While every effort is made to provide dependable information, the publisher, author, and editors cannot be held responsible for any errors or omissions.

Printed in the United States of America.

Distributed by Prentice Hall
A division of Simon & Schuster
Englewood Cliffs, NJ 07632

Prentice Hall International (UK) Limited, *London*
Prentice Hall of Australia Pty. Limited, *Sydney*
Editora Prentice Hall do Brasil, Ltda., *Rio de Janeiro*
Prentice Hall Canada, Inc., *Toronto*
Prentice Hall Hispanoamericana, S.A., *Mexico*
Prentice Hall of India Private Limited, *New Delhi*
Prentice Hall of Japan, Inc., *Tokyo*
Prentice Hall of Southeast Asia Pte. Ltd., *Singapore*

Acknowledgements

The information contained in this book has been obtained from a wide variety of authorities who are specialists in their respective fields. Appreciation is expressed to all those who have contributed their expertise to this volume. Many of the chapters in this volume were originally presented at the World Energy Engineering Congress sponsored by the Association of Energy Engineers.

Contents

SECTION I: Fundamentals of HVAC
- Chapter 1 Heating, Ventilating, Air Conditioning and Building System Optimization 1

SECTION II: Heat Pumps
- Chapter 2 Industrial and Commercial Heat Pump Applications 43
- Chapter 3 The Earth-Coupled Heat Pump As a Demand-Side Planning Tool 69
- Chapter 4 *Case Study:* Heat Pump Heat Recovery System At Tend-R-Fresh 85
- Chapter 5 *Cast Study:* Heat Pump Strategy At the Nevada Test Site 97

SECTION III: Thermal Storage
- Chapter 6 Thermal Storage Options for HVAC Systems 107
- Chapter 7 Analysis of Thermal Storage Options 121
- Chapter 8 *Case Study:* Three Proposed Cold Thermal Storage Systems for a U.S. Navy Shore Facility 143
- Chapter 9 *Case Study:* Thermal Energy Storage for Municipal Buildings 167

SECTION IV: Evaporative Cooling
- Chapter 10 Survey of the Indirect Evaporative Cooling Field 191
- Chapter 11 Theory Vs. Practice In Evaporative Roof Spray Cooling 217

SECTION V: HVAC Controls
- Chapter 12 Comparison of Cost and Performance Of HVAC Controls 241
- Chapter 13 Distributed Intelligence and Communication in Building Automation Systems 261

SECTION VI: Cooling Tower Optimization
- Chapter 14 Cooling Tower Optimization — Innovations In Cooling Tower Heat Transfer 273
- Chapter 15 Maximizing Cooling Tower Potential With State of The Art 293

SECTION VII: HVAC Design Considerations
- Chapter 16 Energy Effects of ASHRAE Standard 62-1981 303
- Chapter 17 Can Absorption Chillers Beat
Electric Centrifugal Chillers? 317
- Chapter 18 Humidification Steam Vs. Water. 333
- Chapter 19 Enthalpy Vs. Dry Bulb Start Time Optimization 343
- Chapter 20 Technology Update for
Desiccant-Based Air Conditioning Systems 349
- Chapter 21 Radiant Heat for Affordable Comfort 365
- Chapter 22 Spreadsheet Models to Determine
HVAC System Savings 377

INDEX 403

SECTION I
FUNDAMENTALS
OF HVAC

1

Heating, Ventilating, Air Conditioning, and Building System Optimization

This chapter will review the basics of Heating, Ventilation and Air Conditioning (HVAC) and buildings as related to energy engineering.

DEGREE DAYS

Degree days are the summation of the product of the difference in temperature (ΔT) between the *average outdoor* and hypothetical *average indoor* temperatures (65°F), and the number of days (t) the outdoor temperature is below 65°F. Therefore:

$$DD\text{'s} = \Delta T \times t, \text{ therefore } \Delta t = DD/t \qquad \textit{Formula (1-1)}$$

Degree Days divided by the total number of days on which Degree Days were accumulated will yield an average ΔT for the season, based on an assumed indoor temperature of 65°F. To find the average outdoor temperature of the season, this figure must be subtracted from 65°F.

Example Problem 1-1

If there are 6750 degree days recorded over a heating season of 270 days, what is the mean outdoor temperature for that season?

Answer

$$\Delta T = DD/t \quad \Delta T = \frac{6740 DD}{270 \text{ days}} \quad \Delta T = 25°F$$

The average outdoor temperature can now be found, since

$\Delta T = T$ (avg. indoor) $- T$ (avg. outdoor)
T (avg. outdoor) $= T$ (avg. indoor) $- \Delta T$
T (avg. outdoor) $= 65°F - 25°F$
T (avg. outdoor) $= 40°F$

RESISTANCE (R) TO HEAT FLOW AND CONDUCTANCE (U) AND CONDUCTIVITY (K)

The rate at which heat flows through a material depends on its characteristics. Some materials transmit heat more readily than others. This characteristic of materials which affects the flow of heat through them, can be viewed either as their *resistance* to the flow of heat or as their *conductance* allowing the flow of heat.

For a section of a building, such as a wall, the conductance is expressed as the U-value for that wall; that is, the number of Btu's that will pass through a one-square-foot section of a building in one hour with a one-degree temperature difference between the two surfaces.

U = Btu's per square foot per hour per degree Fahrenheit.
or
$U = Btu/ft^2 h °F$

R-Value = Thermal Resistance = The unit time for a unit area of a particular body or assembly having defined surfaces with a unit average temperature difference established between the two surfaces per unit of Thermal Transmission.

$$\frac{hr \cdot ft^2 \cdot °F}{Btu}$$

$R = 1/U$ *Formula (1-2)*

The conductivity of a material as related to conductance and resistance is illustrated by Formula 7-3.

$$U = \frac{K}{d} = \frac{1}{R} \qquad Formula\ (1\text{-}3)$$

where d is the thickness of the material.

VOLUME *(V)* OF AIR

The volume of air within a structure is constant even though the air itself changes—new air enters and old air leaves. The total volume is equal to the volume of space within the conditioned portion of the home. (Only the volume of conditioned space is considered since air entering and leaving the unconditioned part of the home does not demand energy to condition it.)

To determine the volume *(V)* of air, multiply the height *(H)* of the space times the width *(W)* of the space times the length *(L)* of the space.* While this can be done for the home as a whole, it is more accurate to calculate it for each room and then add these volumes.

AIR CHANGES PER HOUR *(AC)*

The rate at which the volume of air in a structure changes per hour differs greatly from building to building. The number of air changes per hour *(AC/*h) has wide variation due to a number of factors such as

- *The number and size of openings* in the envelope—around doors and windows and in the siding itself;
- *The average speed* of the wind blowing against the structure and the protection the structure has from this wind;
- *The number and size* of chimneys, vents, and exhaust fans and the frequency of their use;
- *The number of times* that doors and windows are opened; and
- *How the structure is used.*

---
*This is only appropriate for structures with flat ceilings.

HEATING CAPACITY OF AIR *(HC)*

Air can be heated and cooled. A certain amount of heat is necessary to change the temperature of each cubic foot (ft^3) of air one degree Fahrenheit (F). This amount of heat depends on the density of air which varies with temperature and pressure. This figure will generally be within the range of 0.018–0.022 Btu's/ft^3 °F.

BUILDING DYNAMICS

The building experiences heat gains and heat losses depending on whether the cooling or heating system is present, as illustrated in Figures 1-1 and 1-2. Only when the total season is considered in conjunction with lighting and heating, ventilation and air-conditioning (HVAC) can the energy utilization choice be decided. One way of reducing energy consumption of HVAC equipment is to reduce the overall heat gain or heat loss of a building.

CONDUCTION HEAT LOSS

The formula used to determine the amount of heat conducted through the envelope is as follows: Degree days *(DD)* is the product of the difference in temperature, ΔT, and the time *(t)* in days, providing that the days in degree days *(DD)* are converted to hours. This is accomplished by multiplying *(DD)* times 24 hours a day. This will yield the quantity of heat *(Q)* conducted through a particular section of the envelope for the entire heating season.

The formula can be written:

$$Q_{\text{(heating season)}} = U \times A \times DD \times 24 \text{ hours/day} \qquad \text{Formula (1-4)}$$

or

$$Q_{\text{(heating season)}} = \frac{A \times DD \times 24 \text{ hrs/day}}{R} \qquad \text{Formula (1-5)}$$

In general heat flow through a flat surface is defined as

$$Q = U A \Delta T \qquad \text{Formula (1-6)}$$

where ΔT is the temperature difference causing the heat flow.

Heating, Ventilating, Air Conditioning, and Building System Optimization 5

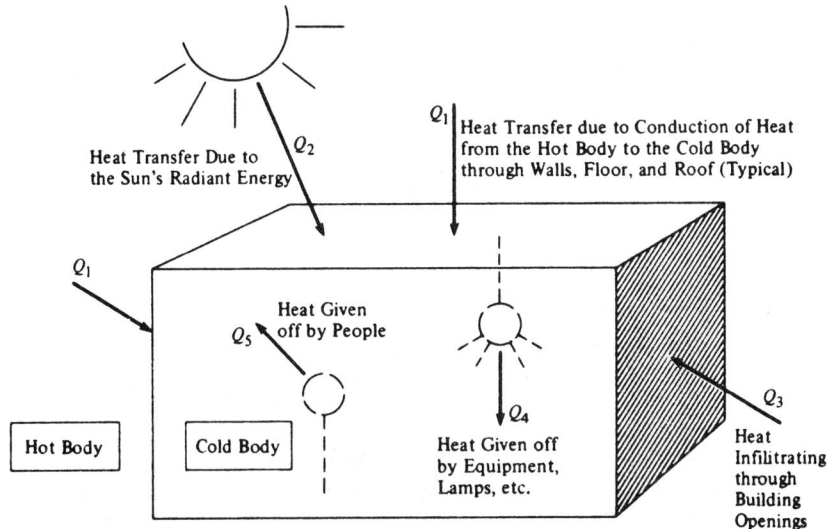

Heat Gain = $Q_1 + Q_2 + Q_3 + Q_4 + Q_5$

Figure 1-1. Heat Gain of a Building

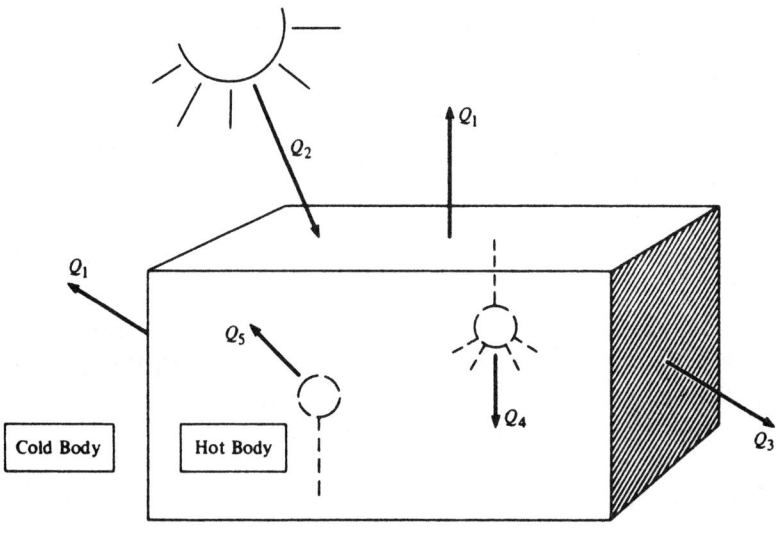

Heat Loss = $Q_1 + Q_3 - Q_2 - Q_4 - Q_5$

Figure 1-2. Heat Loss of a Building

For a composite wall, the heat flow is represented by Figure 1-3. To calculate the overall U or Conductance value, the resistance of each material is added in series. This is analogous to an electrical circuit.

$$R = R_1 + R_2 + R_3 + R_4 + R_5 \qquad Formula\ (1\text{-}7)$$

FACTORS IN CONDUCTION

There are four factors which affect the conduction of heat from one area to another. They are:

- *The difference in temperature (ΔT)* between the warmer area and the colder area;
- *The length of time (t)* over which the transfer occurs;
- *The area (A) in common* between the warmer and the colder area; and
- *The resistance (R) to heat flow and conduction (U)* between the warmer and the colder area.

Difference in Temperature

Heat flows (much as water moves down hill) from warm areas to cold ones. The steeper the gradient between its origin and its destination the faster it will flow. In fact, the rate at which heat is conducted is directly proportionate to the difference in temperature (ΔT) between the warm area and the colder one.

Length of Time

The longer the heat is allowed to flow across the gradient, the more heat will be conducted. The amount of heat (Btu's) is directly proportionate to the time span *(t)* of the transfer.

Btu/h is the amount of heat transferred in one hour.

The Area *(A)* in Common

The larger the area common to the warmer and colder surfaces, through which the heat flows, the greater is the rate of conducted heat. For the same material, for the same length of time, at the same ΔT, the amount of heat (Btu's) transferred is directly proportionate to the area (A) in common.

Example Problem 1-2

Calculate the heat loss through 20,000 ft² of building wall, as indicated by Figure 1-3. Assume a temperature differential of 17°F.

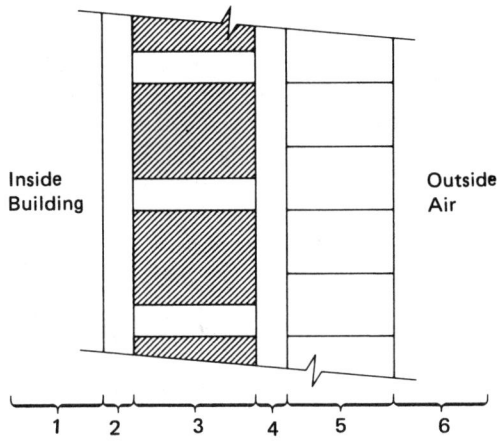

1 Inside Air Film $R = .68$
2 ½" Plaster Board Interior Finish $R = .44$
3 8" Concrete Block, Sand & Gravel Aggregate $R = 1.1$
4 ½" Cement Mortar $R = .1$
5 4" Brick Exterior $R = .44$
6 Outside Air Film @ 15 mph Wind $R = .17$

Figure 1-3. Typical Wall Construction

Answer

Description	Resistance
Outside air film at 15 mph	0.17
4" brick	0.44
Mortar	0.10
Block	1.11
Plaster Board	0.44
Inside film	0.68
Total resistance	2.94

$U = 1/R = 0.34$
$Q = U A \Delta T$
$ = 0.34 \times 20{,}000 \times 17 = 115{,}600$ Btu/h.

In Figure 1-4 a surface film conductance is introduced.

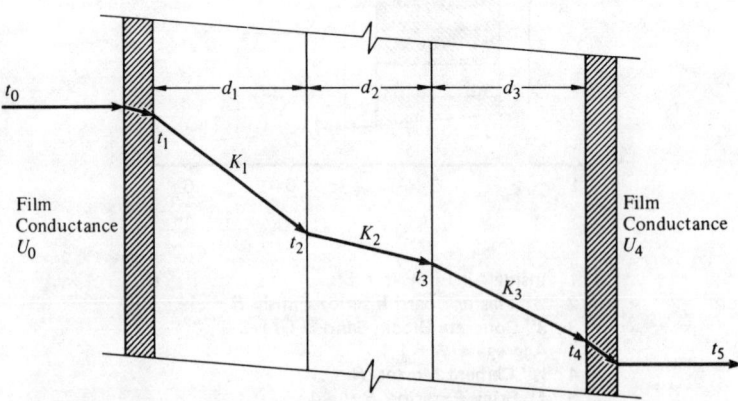

Figure 1-4. Temperature Distribution for the Composite Wall

The surface or film conductance is the amount of heat transferred in Btu per hour from a surface to air or from air to a surface per square foot for one degree difference in temperature. The flow of heat for the composite material can also be specified in terms of the conductivity of the material and the conductance of the air film:

$$Q = \frac{A(t_0 - t_5)}{1/U_0 + d_1/K_1 + d_2/K_2 + d_3/K_3 + 1/U_4}.$$ Formula (1-8)

LATENT HEAT AND SENSIBLE HEAT

The latent heat gain of a space means that moisture is being added to the air in the space. Moisture in the air is really in the form of superheated steam. Removing sensible heat from a space through air-conditioning equipment lowers the dry-bulb temperature of the air. On the other hand, removing latent heat from a space changes the substance state from a vapor to a liquid. The latent heat gain of a space is expressed in terms of moisture, heat units (Btu) or grains of moisture per hour (7,000 grains equals one pound). The average value for the latent heat or vaporization for superheated steam in air is 1050.

Example Problem 1-3

2000 grains of moisture are released in a conditioned room each hour. Calculate the heat that must be removed in order to condense this moisture at the cooling coils.

Answer

$$\frac{2000}{7000} \times 1050 = 299.9 \text{ Btuh}$$

Example Problem 1-4

Calculate the quantity of heat (Q) required by infiltration in an 8,000 cubic foot (ft^3) home that has 1.7 air changes per hour (AC/h) when the outside temperature is 48°F and the inside temperature is 68°F (ΔT = 20°F) for one day (24 hours).

Answer

$Q = V \times AC/h \times 0.020^* \text{ Btu/ft}^3 \, ^\circ F \times \Delta T \times t$
$Q = 8{,}000 \text{ ft}^3 \times 1.7\, AC/h \times 0.020^* \text{ Btu/ft}^3 \, ^\circ F \times 20^\circ F \times 24 \text{ hrs}$
$Q = 8{,}000 \times 1.7 \times 0.020^* \times 20 \times 24$
$Q = 130{,}560 \text{ Btu's}^{**}$

INFILTRATION

Leakage or infiltration of air into a building is similar to the effect of additional ventilation. Unlike ventilation, it cannot be controlled or turned off at night. It is the result of cracks, openings around windows and doors, and access openings. Infiltration is also induced into the building to replace exhaust air unless the HVAC balances the exhaust. Wind velocity increases infiltration and stack effects are potential problems. Air that is pushed out the window and door cracks is referred to as exfiltration.

To estimate infiltration, the Air Change Method or Crack Method is used.

Air Change Method

The five factors which determine the amount of energy lost through infiltration can be put together in a formula that states that:

The quantity of heat (Q) equals the volume of air (V) times the number of air changes per hour (AC/h) times the amount of heat required to raise the temperature of air one degree Fahrenheit (0.018–0.022 Btu's) times the temperature difference (ΔT) times the length of time (t). This is expressed as follows:

$Q = V \times AC/h \times 0.020 \text{ Btu's} \cdot /\text{ft}^3 \, ^\circ F \times \Delta T \times t$ *Formula (1-9)*

The Air Change Method is considered to be a quick estimation method and is not usually accurate enough for air-conditioning design. A second method used to determine infiltration is the Crack Method.

---
*This is a regional variable.
** Once again, all of the units in the formula cancel except Btu's, leaving the units for Q as Btu's.

Crack Method

When infiltration enters a space it adds sensible and latent heat to the room load. To calculate this gain the following equations (1-10) and (1-11) are used.

Sensible Heat Gain

$$Q_S = 1.08 \text{ CFM } \Delta T \qquad \textit{Formula (1-10)}$$

Latent Heat Gain

$$Q_L = .7 \text{ CFM } (HR_o - HR_i) \qquad \textit{Formula (1-11)}$$

where
- Q_S = Sensible heat gain Btuh
- Q_L = Latent heat gain Btuh
- CFM = Air Flow Rate
- ΔT = Temperature differential between outside and inside air, F
- HR_o = Humidity ratio of outside air, grains per lb
- HR_i = Humidity ratio of room air, grains per lb

BODY HEAT

The human body releases sensible and latent heat depending on the degree of activity. Heat gains for typical applications are summarized in Table 1-1.

EQUIPMENT, LIGHTING AND MOTOR HEAT GAINS

It is important to include heat gains from equipment, lighting systems and motor heat gain in the overall calculations.

For a manufacturing facility, the major source of heat gains will be from the process equipment. Consideration must be given to all equipment including motors driving supply and exhaust fans.

To convert motor horsepower to heat gain in Btuh, equation 1-12 is used.

Table 1-1. Heat Gain from Occupants

Activity	Sensible Heat Btuh	Latent Heat Btuh	Total Heat Gain Btuh
Very Light Work — Seated (Offices, Hotels, Apartments)	215	185	400
Moderately Active Work (Offices, Hotels, Apartments)	220	230	450
Moderately Heavy Work (Manufacturing)	330	670	1000
Heavy Work (Manufacturing)	510	940	1450

Source: ASHRAE—Guide & Data Book

$$Q = \frac{hp \times .746}{\eta} \times 3412 \qquad Formula\ (1\text{-}12)$$

where
- hp is the running motor horsepower
- η is the efficiency of the motor
- Q is the heat gain from the motor Btuh

Similarly, the kilowatts of the lighting system can be converted to heat gain.

$$Q = (KW_F + KW_B) \times 3412 \qquad Formula\ (1\text{-}13)$$

where
- KW_F is the kilowatts of the lighting fixtures
- KW_B is the kilowatts of the ballast
- Q is the heat gain from the lighting system Btuh

RADIANT HEAT GAIN

Heat from the sun's rays greatly increases heat gain of a building. If the building energy requirements were mainly due to cooling, then this gain should be minimized. Solar energy affects a building in the following ways:

1. *Raises the surface temperature:* Thus a greater temperature differential will exist at roofs than at walls.

2. A large percentage of direct solar radiation and diffuse sky radiation *passes through* transparent materials, such as glass.

SURFACE TEMPERATURES

The temperature of a wall or roof depends upon:

(a) the angle of the sun's rays
(b) the color and roughness of the surface
(c) the reflectivity of the surface
(d) the type of construction.

When an engineer is specifying building materials, he should consider the above factors. A simple example is color. The darker the surface, the more solar radiation will be absorbed. Obviously, white surfaces have a lower temperature than black surfaces after the same period of solar heating. Another factor is that smooth surfaces reflect more radiant heat than do rough ones.

In order to properly take solar energy into account, the angle of the sun's rays must be known. If the latitude of the plant is known, the angle can be determined.

SUNLIGHT AND GLASS CONSIDERATIONS

A danger in the energy conservation movement is to take steps backward. A simple example would be to exclude glass from building designs because of the poor conductance and solar heat gain factors of clear glass. The engineer needs to evaluate various alternate glass constructions and coatings in order to maintain and improve the aesthetic qualities of good design while minimizing energy inefficiencies. It should be noted that the method to reduce heat gain of glass due to conductance is to provide an insulating air space.

To reduce the solar radiation that passes through glass, several techniques are available. Heat absorbing glass (tinted glass) is very popular. Reflective glass is gaining popularity, as it greatly reduces solar heat gains.

To calculate the relative heat gain through glass, a simple method is illustrated below:

$$Q = \pm UA(t_0 - t_1) + A \times S_1 \times S_2 \qquad \text{Formula (1-14)}$$

where
- Q is the total heat gain for each glass orientation (Btuh)
- U is the conductance of the glass (Btu/h-ft^2-°F)
- A is the area of glass; the area used should include framing, since it will generally have a poor conductance compared with the surrounding material. (ft^2)
- $t_0 - t_1$ is the temperature difference between the inside temperature and outside ambient. (°F)
- S_1 is the shading coefficient; S_1 takes into account external shades, such as venetian blinds and draperies, and the qualities of the glass, such as tinting and reflective coatings.
- S_2 is the solar heat gain factor. This factor takes into account direct and diffused radiation from the sun. Diffused radiation is basically caused by reflections from dust particles and moisture in the air.

THE PSYCHROMETRIC CHART

Just as the steam table and the Mollier Diagram are used to relate the properties of steam, the psychrometric chart is used to illustrate the properties of air. The psychrometric chart is a very important tool in the design of air-conditioning systems.

PROPERTIES OF AIR

Air expands and contracts with temperature. If pressure is held constant, then air expands or contracts at a specified rate with change in temperature as defined by equation 1-15.

$$V_2 = V_1 \frac{T_2}{T_1} \qquad \text{Formula (1-15)}$$

where

P_2 = initial pressure, psia
P_1 = final pressure, psia

The change in the volume occupied by air at any temperature can be found by first using Formula 1-15 to calculate the change in volume with pressure and then using Formula 1-14 to calculate the change in volume with temperature.

The temperature at which the water vapor in the atmosphere begins to condense is the *Dew Point Temperature.* It should be noted that the weight of moisture per pound of dry air in a mixture of air and water vapor depends on the dew point temperature alone. If there is no condensation of moisture, the dew point temperature remains constant.

Humidity Ratio is defined as the weight of water vapor mixed with one pound of dry air.

Degree of Saturation is defined as the actual humidity ratio divided by the humidity ratio at saturation.

Relative Humidity is defined as the vapor pressure of air divided by the saturation pressure of pure water at the same temperature.

Sensible Heat of an Air-Vapor Mixture is defined as the heat which affects the dry-bulb temperature of the mixture only.

Wet-Bulb Temperature can be determined by covering the bulb of a thermometer with a wet wick and holding it in a stream of swiftly moving air. At first the temperature will drop quickly and then reach a stationary point referred to as the wet-bulb temperature. The wet-bulb temperature is lower than the dry thermometer. The amount of water which evaporates from the wet wick into the air depends on the amount of water vapor initially in the air flowing past the wet bulb. A sling psychrometer is a convenient instrument used to measure wet-bulb temperatures.

Total Heat is defined as the sum of the sensible and latent heat. Sensible heat depends only upon the dry-bulb temperature, while the latent heat content depends only upon its dew point.

The *Enthalpy of an Air-Water Vapor Mixture* can be calculated by equation 1-16.

$$h_{(mix)} = h_{(dry\ air)} + h_{(water\ vapor)} \qquad Formula\ (1\text{-}16)$$

or

$$h_{(mix)} = Cp \times T_{DB} + HR + hg$$

where
- $h_{(mix)}$ = enthalpy of the mixture of dry air and water vapor, Btu per lb
- Cp = specified heat, Btu per LvF
- T_{DB} = dry bulb temperature
- HR = humidity ratio of the mixture
- hg = enthalpy of saturated vapor (steam) at the dew point temperature

To determine the properties of air such as the humidity ratio, relative humidity, enthalpy, the psychrometric chart is frequently used. Figure 7-4 illustrates the psychrometric chart.

Example Problem 1-4

Given air at 70° DB and 50% relative humidity, for the air vapor mixture find:
- Wet-Bulb Temperature
- Enthalpy
- Humidity Ratio
- Dew Point Temperature
- Specific Volume
- Vapor Pressure
- Percentage Humidity

Answer

From Figure 1-4 find the intersection of 70° DB and 50% *HR*, point "A."

The WB temperature is found as 58.6 WB, point "B."
The Enthalpy is found as 25.5 Btu/lb, point "C."

Heating, Ventilating, Air Conditioning, and Building System Optimization 17

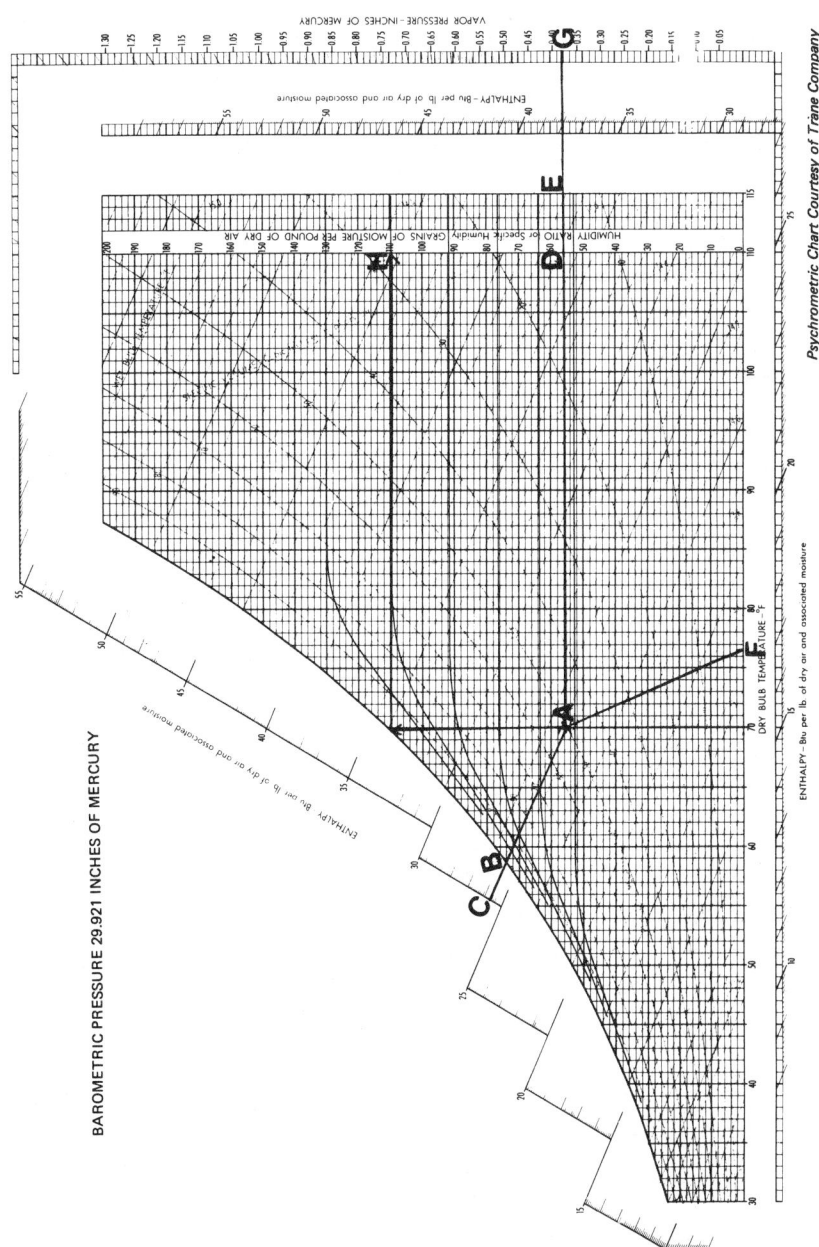

Figure 1-4. Psychrometric Chart

The Humidity Ratio is found to be 56 grains of moisture per pound of dry air, point "D" and the Dew Point temperature is 53°F, point "E."

The Specific Volume is found to be 13.5 cubic feet per pound of air, point "F," with a Vapor Pressure of .38 inches of mercury, point "G."

The percentage humidity equals the actual humidity 56, point "D" divided by the humidity ratio at saturation (100% RH) which is found to be 110, point "H." Thus % humidity = 56/110 = .50.

Example Problem 1-5

Given 8000 CFM of chilled air at 55°F DB and 50°F WB mixed with 3000 CFM of outside air at 90°F DB and 80°F WB, compute the properties of the mixture.

Answer

From Figure 1-5, the intersection of 55°F DB and 50°F WB is point "A." The specific volume is then 13.1 cubic feet/lb, point "B."

Similarly, for the outside air, the specific volume is 14.3 cubic feet/lb, point "D."

The total weight and dry-bulb temperature of the mixture can be found by the following ratios:

$$\frac{8,000}{13.1} = 610.6 \text{ lb/min.}$$

$$\frac{3,000}{14.3} = 209.7 \text{ lb/min.}$$

$$\overline{820.3 \text{ lb/min. for total weight}}$$

The dry-bulb temperature is:

$$\frac{610.6}{820.3} \times 55 = 40.93°F$$

Heating, Ventilating, Air Conditioning, and Building System Optimization

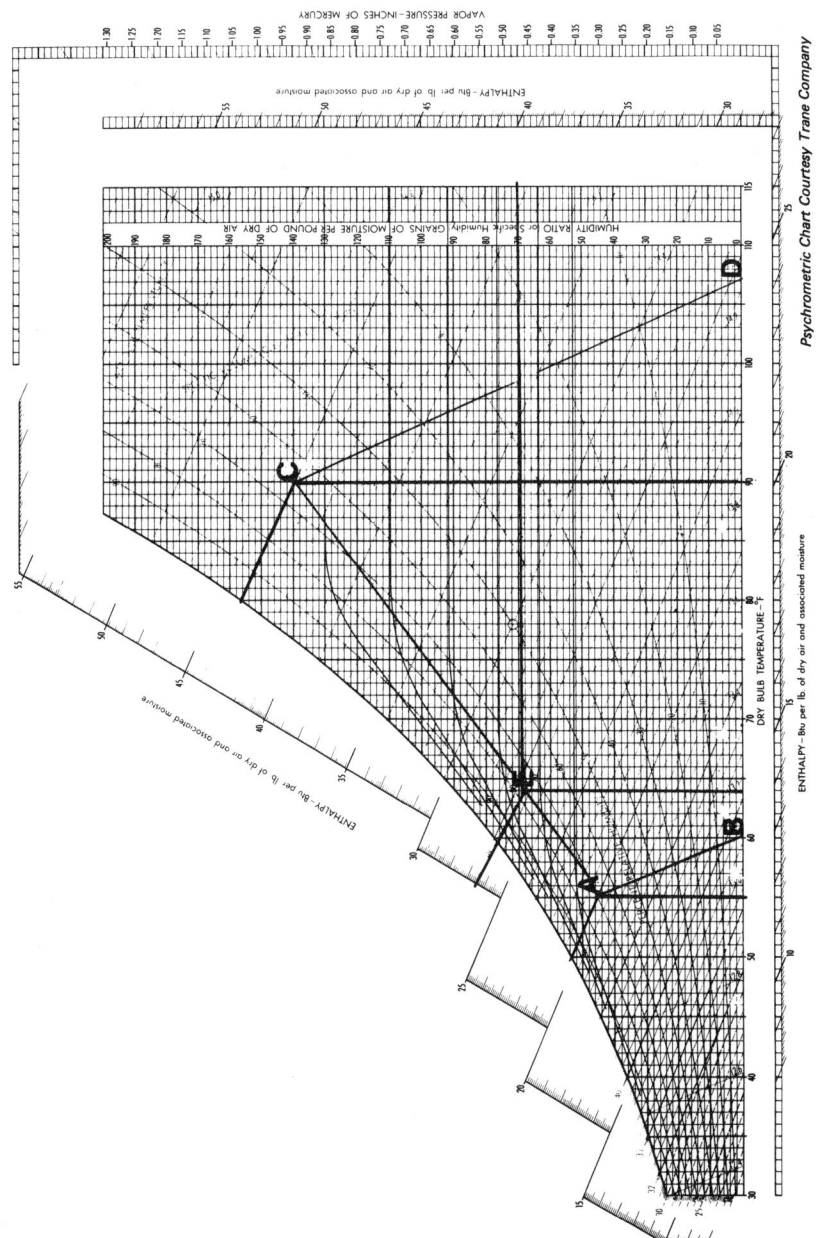

Figure 1-5. Psychrometric Chart for Mixture of Air

$$\frac{209.7}{820.3} \times 90 = \underline{23.0°F}$$
$$\underline{63.9°F \text{ DB}}$$

The properties of the mixture, point "E" can now be determined from the chart.

WB = 59.8°F
h = 26.6 Btu/lb
humidity ratio = 70 gr of moisture/lb of dry air.

BASICS OF FAN DISTRIBUTION SYSTEMS

In order to distribute conditioned or ventilated air, fans are the chief vehicle used. Several basic types of fans commonly used in industry are illustrated in Figure 1-6 and are listed below:

Centrifugal, airfoil blade—used on large heating, ventilating, and air-conditioning systems. Airfoil fans are used where clean air is handled.

Centrifugal, backward curved blade—used for general heating, ventilating, and air-conditioning systems. Air handled need not be as clean as above.

Centrifugal, radial blade—a rugged, heavy duty fan for high pressure applications. It is designed to handle sand, wood chips, etc.

Centrifugal, forward curved blade—ideal for low pressure applications, such as domestic furnaces or room and packaged air conditioners.

The brake horsepower of a fan is illustrated by Formula 1-17.

$$\text{Brake Horsepower} = \frac{\text{CFM} \times \text{Fan PS}}{6356 \times \eta_F} \qquad \textit{Formula (1-17)}$$

where
CFM is the quantity of air in CFM
η is the fan static efficiency
Fan PS is the Fan Static Pressure in inches

To compute the fan static pressure

Heating, Ventilating, Air Conditioning, and Building System Optimization 21

Figure 1-6. Fan Types: (A) Vaneaxial; (B) Backward Curved Blade;
(C) Tubeaxial; (D) Radial; (E) Radial Tip Blade; (F) Airfoil Blade
Courtesy of Buffalo Forge Company

$$\text{Fan PS} = P_T(O) - P_T(i) - P_v(O) \qquad \textit{Formula (1-18)}$$

where

    $P_T(O)$ is the total pressure at fan outlet
    $P_T(i)$ is the total pressure at fan inlet
    $P_v(O)$ is the velocity pressures at fan outlet

The excess pressure above the static pressure is known as the velocity pressure, and is computed by Formula 7-19 for standard air having a value of 13.33 cu ft per lb as

$$PV = \left(\frac{V}{4005}\right)^2 \qquad \textit{Formula (1-19)}$$

where V is the velocity of air in FPM.

Note that the pressure of air in sheet metal ducts is so low that ordinary pressure gauges (Bourdon type) cannot be used, thus a V-tube or manometer is used, which measures pressure in inches of water. A pressure of 1 psi will support a column of water 2.31 ft high or 27.7 inches.

Fan Laws

The performance of a fan at varying speeds and air densities may be predicted by certain basic fan laws as illustrated in Table 1-2.

Example Problem 1-6

An energy audit indicates that the ventilation requirements of a space can be reduced from 15,000 CFM to 12,000 CFM. Comment on the savings in brake horsepower if the fan pulley is changed to reduce the fan speed accordingly.

Answer

From the Fan Laws:

Table 1-2. Fan Laws

1.	Fan Law for variation in fan speed at constant air density with a constant system	
	1.1	Air volume, CFM varies as fan speed
	1.2	Static velocity or total pressure varies as the square of fan speed
	1.3	Power varies as cube of fan speed
2.	Fan Law for variation in air density at constant fan speed with a constant system	
	2.1	Air volume is constant
	2.2	Static velocity or pressure varies as density
	2.3	Power varies as density

$$hp_1 = hp_2 \times \left(\frac{\text{CFM new}}{\text{CFM old}}\right)^3$$

$$= hp_2 \times \left(\frac{12{,}000}{15{,}000}\right)^3 = hp_2 (.8)^3 = .512\, hp_2$$

or a 48.8% savings.

The fan performance is affected by the density of the air that the fan is handling. All fans are rated at standard air with a density of .075 lb per cu ft and a specific volume of 13.33 cu ft per lb. When a fan is tested in a laboratory at different than standard air, the brake horsepower is corrected by using the Fan Laws.

Fan Performance Curves

Fan performance curves are used to determine the relationship between the quantity of air that a fan will deliver and the pressure it can discharge at various air quantities. For each fan type, the manufacturer can supply fan performance curves which can be used in design and as a tool of determining the fan efficiency.

As illustrated by problem 1-6, one energy engineering technique to reduce fan horsepower is to reduce fan speed. An alternate way is to throttle the air flow by a damper. The fan performance curves can be used to illustrate the best choice of these options. The system characteristics can be plotted on the fan curves to show the static pressure required to overcome the friction loss in the duct system. From the Fan Laws the system friction loss varies with the square of fan speed; thus, as the air quantity increases, the friction loss will vary as illustrated by Figure 1-7.

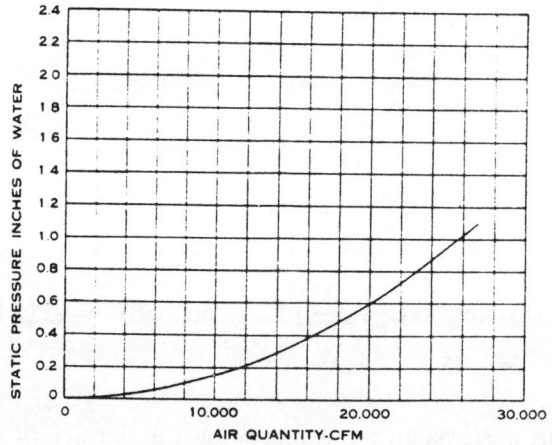

Figure 1-7. System Characteristic Curve

For a detailed analysis, the fan performance curve should be used to predict how a specific fan will perform in a desired application.

Example Problem 1-7

Given Fan Performance Curve Figure 7-8. The fan delivers 21,500 CFM at 600 rpm at a brake horsepower of 12.3. Comment on the savings in brake horsepower by reducing air flow to 14,400 CFM by each of the following methods: (a) reducing fan speed to 400 rpm, (b) throttling the air flow by a damper.

Heating, Ventilating, Air Conditioning, and Building System Optimization

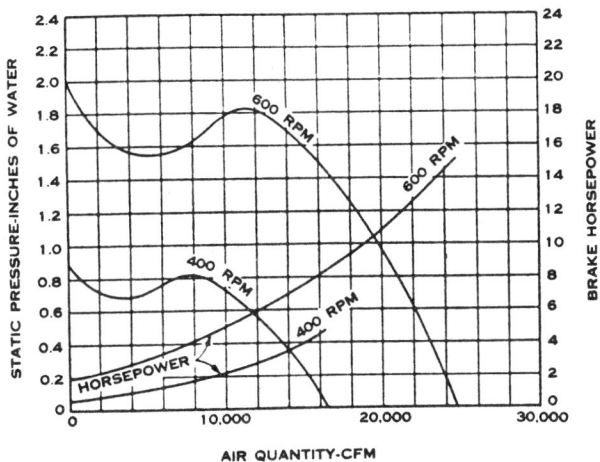

Figure 1-8. Fan Performance Curves

(a) From the Fan Laws the brake horsepower is reduced as follows:

$$hp = (12.3)\left(\frac{400}{600}\right)^3 = 3.64$$

Using the fan performance curve, Figure 7-9, the system characteristic curve "A" is plotted.

By reducing the rpm from 600 to 400, the system operates at point 1 and then moves to point 2. The brake horsepower is found from Figure 1-8 to be 3.7.

(b) By closing the air damper, the air flow is reduced to 14,400 CFM while still running the fan at 600 rpm. Using the fan performance curve Figure 1-9, the system operates at point 1 and then moves to point 3. The power to operate the fan at point 3 is 7.2 hp from Figure 1-8. Thus, if the fan speed can be reduced it is more efficient than throttling the air flow damper.

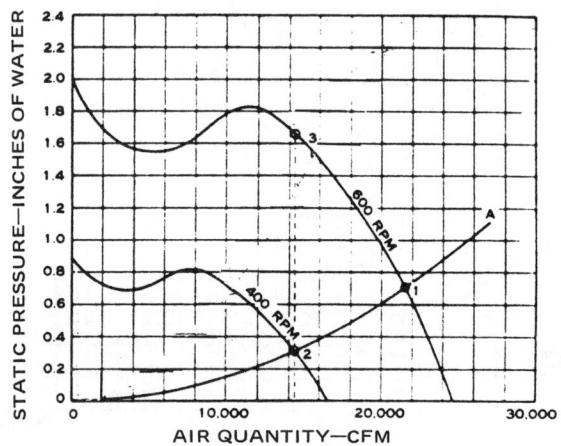

Figure 1-9. Fan Performance and System Characteristic Curves

FLUID FLOW

Pump and piping considerations are extremely important due to the fact that energy transport losses are a part of any distribution system. Losses occur due to friction and that lost energy must be supplied by pump horsepower.

Centrifugal pumps are commonly used in heating, ventilation and air-conditioning applications as well as utility systems. The output torque for the pump is supplied by a driver such as a motor. Liquid enters the eye of the impeller which rotates. Pressure energy builds up by the action of centrifugal force, which is a function of the impeller vane peripheral velocity.

As with fan systems, Pump Laws and curves can be used to predict system responsiveness. The affinity laws of a pump are illustrated in Table 1-3.

The horsepower required to operate a pump is illustrated by Formula 1-20.

$$\text{hp} = \frac{\Delta P \text{ GPM}}{17.5 \eta} \qquad \textit{Formula (1-20)}$$

Table 1-3. Affinity Laws for Pumps

Impeller Diameter	Speed	Specific Gravity (SG)	To Correct for	Multiply by
Constant	Variable	Constant	Flow	$\left(\dfrac{\text{New Speed}}{\text{Old Speed}}\right)$
			Head	$\left(\dfrac{\text{New Speed}}{\text{Old Speed}}\right)^2$
			BHP (or kW)	$\left(\dfrac{\text{New Speed}}{\text{Old Speed}}\right)^3$
Variable	Constant		Flow	$\left(\dfrac{\text{New Diameter}}{\text{Old Diameter}}\right)$
			Head	$\left(\dfrac{\text{New Diameter}}{\text{Old Diameter}}\right)^2$
			BHP (or kW)	$\left(\dfrac{\text{New Diameter}}{\text{Old Diameter}}\right)^3$
Constant		Variable	BHP (or kW)	$\dfrac{\text{New SG}}{\text{Old SG}}$

where

ΔP is the differential pressure across a pump in psi
GPM is the required flow rate in gallons per minute
η is the pump efficiency

To convert psi to read in feet use Formula 1-21.

$$\text{Head in Feet} = \frac{\text{psi} \times 2.31}{\text{Specific Gravity of Fluid}} \qquad \textit{Formula (1-21)}$$

Basically, the size of discharge line piping from the pump determines the friction loss through the pipe that the pump must overcome. The greater the line loss, the more pump horsepower required. If the line is short or has a small flow, this loss may not be significant in terms of the total system head requirements. On the other hand, if the line is long and has a large flow rate, the line loss will be significant.

To calculate the pressure loss for water system piping and the corresponding velocity, Formulas 1-22 and 1-23 are used.

$$\Delta P = \frac{.055 \, CF^{1.85}}{d^{4.87}} \qquad \text{Formula (1-22)}$$

$$V = \frac{.41 \, F}{d^2} \qquad \text{Formula (1-23)}$$

where
- ΔP = pressure loss per 100 feet of pipe, psi
- V = velocity of fluid, ft/sec.
- C = roughness factor
  - 1 for copper tubing
  - 1.62 for steel pipe
  - .77 for plastic pipe
- F = flow rate in gallons per minute
- d = inside diameter of pipe, inches

The pressure loss due to fittings is determined by Formula 1-24.

$$\Delta P = .0067 \, KV^2 \qquad \text{Formula (1-24)}$$

where K is the loss coefficient.

Options to Reduce Pump Horsepower

There are several alternates that will significantly reduce pump horsepower. The below summarizes some of the options available.

1. Many pumps are oversized due to very conservative design practices. If the pump is oversized, install a smaller impeller to match the load.

2. In some instances heating or cooling supply flow rates can be reduced. To save on pump horsepower either reduce motor speed or change the size of the motor sheave.
3. Check economics of replacing corroded pipe with a large pipe diameter to reduce friction losses.
4. Consider using variable speed pumps to better match load conditions. Motor drive speed can be varied to match pump flow rate or head requirements.
5. Consider adding a smaller auxiliary pump. During part load situations a larger pump can be shut down and a smaller auxiliary pump used.

HVAC SYSTEMS

The below summary illustrates the types of systems frequently encountered in heating air-conditioning systems.

Single Zone System

Single zone systems consist of a mixing, conditioning and fan section. The conditioning section may have heating, cooling, humidifying or a combination of capabilities. Single zone systems can be factory assembled roof-top units or built up from individual components and may or may not have distributing duct work.

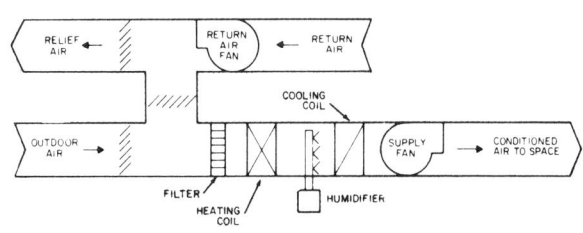

Terminal Reheat System

Reheat systems are modifications of single zone systems. Fixed cold temperature air is supplied by the central conditioning system and reheated in the terminal units when the space cooling load is less than maximum. The reheat is controlled by thermostats located in each condition space.

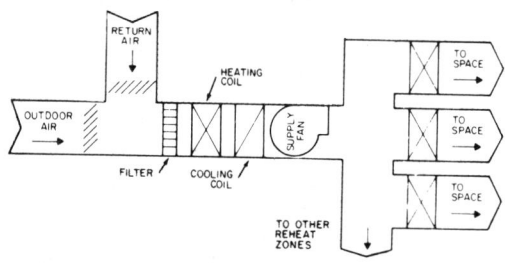

Multizone Systems

Multizone systems condition all air at the central system and mix heated and cooled air at the unit to satisfy various zone loads as sensed by zone thermostats. These systems may be packaged roof-top units or field-fabricated systems.

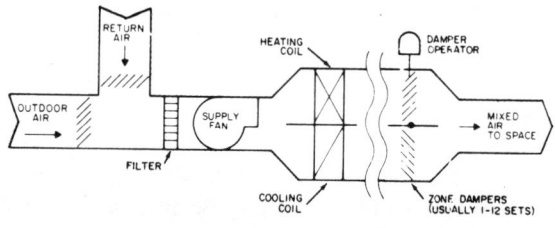

Dual Duct Systems

Dual duct systems are similar to multizone systems except heated and cooled air is ducted to the conditioned spaces and mixed as required in terminal mixing boxes.

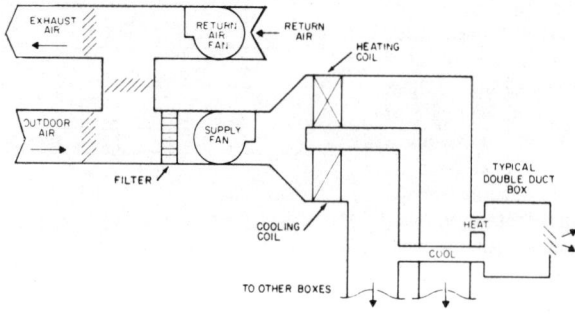

Variable Air Volume Systems

A variable air volume system delivers a varying amount of air as required by the conditioned spaces. The volume control may be by fan inlet (vortex) damper, discharge damper or fan speed control. Terminal sections may be single duct variable volume units with or without reheat, controlled by space thermostats.

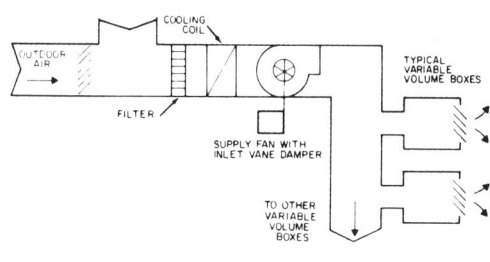

Induction Systems

Induction systems generally have units at the outside perimeter of conditioned spaces. Conditioned primary air is supplied to the units where it passes through nozzles or jets and by induction draws room air through the induction unit coil. Room temperature control is accomplished by modulating water flow through the unit coil.

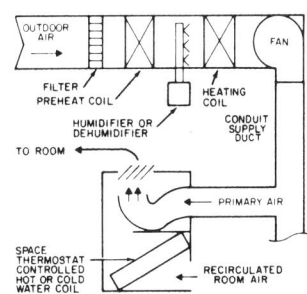

Fan Coil Units

A fan coil unit consists of a cabinet with heating and/or cooling coil, motor and fan and a filter. The unit may be floor or ceiling mounted and uses 100% return air to condition a space.

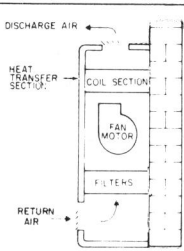

Unit Ventilator

A unit ventilator consists of a cabinet with heating and/or cooling coil, motor and fan, a filter and a return air—outside air mixing section. The unit may be floor or ceiling mounted and uses return and outside air as required by the space.

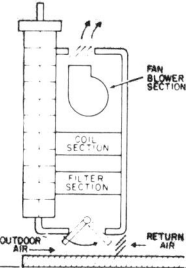

Unit Heater

Unit heaters have a fan and heating coil which may be electric, hot water or steam. They do not have distribution duct work but generally use adjustable air distribution vanes. Unit heaters may be mounted overhead for heating open areas or enclosed in cabinets for heating corridors and vestibules.

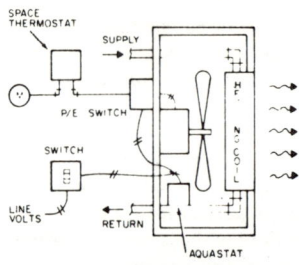

Perimeter Radiation

Perimeter radiation consists of electric resistance heaters or hot water radiators usually within an enclosure but without a fan. They are generally used around the conditioned perimeter of a building in conjunction with other interior systems to overcome heat losses through walls and windows.

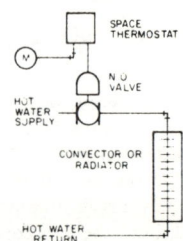

Hot Water Converters

A hot water converter is a heat exchanger that uses steam or hot water to raise the temperature of heating system water. Converters consist of a shell and tubes with the water to be heated circulated through the tubes and the heating steam or hot water circulated in the shell around the tubes.

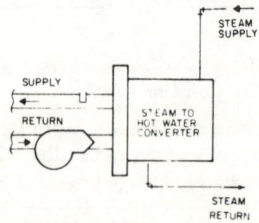

Source: "Energy Conservation with Comfort," Honeywell

Reducing Energy Consumption in HVAC Systems

Variable Air Volume System—A variable volume system provides heated or cooled air at a constant temperature to all zones served. VAV boxes located in each zone or in each space adjust the quantity of air reaching each zone or space depending on its load requirements. Methods for conserving energy consumed by this system include:

1. Reduce the volume of air handled by the system to that point which is minimally satisfactory.
2. Lower hot water temperature and raise chilled water temperature in accordance with space requirements.
3. Lower air supply temperature to that point which will result in the VAV box serving the space with the most extreme load being fully open.
4. Consider installing static pressure controls for more effective regulation of pressure bypass (inlet) dampers.
5. Consider installing fan inlet damper control systems if none now exist.

Constant Volume System—Most constant volume systems either are part of another system—typically dual duct systems—or serve to provide precise air supply at a constant volume. Opportunities for conserving energy consumed by such systems include:

1. Determine the minimum amount of airflow which is satisfactory and reset the constant volume device accordingly.
2. Investigate the possibility of converting the system to variable (step controlled) constant volume operation by adding the necessary controls.

Induction System—Induction systems comprise an air handling unit which supplies heated or cooled primary air at high pressure to induction units located on the outside walls of each space served. The high pressure primary air is discharged within the unit through nozzles inducing room air through a cooling or heating coil in the unit. The resultant mixture of primary air and induced air is discharged to the room at a temperature dependent upon the cooling and heating load of the space. Methods for conserving energy consumed by this system include:

1. Set primary air volume to original design values when adjusting and balancing work is performed.
2. Inspect nozzles. If metal nozzles, common on most older models, are installed, determine if the orifices have become enlarged from years of cleaning. If so, chances are that the volume/pressure relationship of the system has been altered. As a result, the present volume of primary air and the appropriate nozzle pressure required must be determined. Once done, rebalance the primary air system to the new nozzle pressures and adjust individual induction units to maintain airflow temperature. Also, inspect nozzles for cleanliness. Clogged nozzles provide higher resistance to air flow, thus wasting energy.
3. Set induction heating and cooling schedules to minimally acceptable levels.
4. Reduce secondary water temperatures during the heating season.
5. Reduce secondary water flow during maximum heating and cooling periods by pump throttling or, for dual-pump systems, by operating one pump only.
6. Consider manual setting of primary air temperature for heating, instead of automatic reset by outdoor or solar controllers.

Dual-Duct System—The central unit of a dual-duct system provides both heated and cooled air, each at a constant temperature.

Each space is served by two ducts, one carrying hot air, the other carrying cold air. The ducts feed into a mixing box in each space which, by means of dampers, mixes the hot and cold air to achieve that air temperature required to meet load conditions in the space or zone involved. Methods for improving the energy consumption characteristics of this system include:

1. Lower hot deck temperature and raise cold deck temperature.
2. Reduce air flow to all boxes to minimally acceptable level.
3. When no cooling loads are present, close off cold ducts and shut down the cooling system. Reset hot deck according to heating loads and operate as a single duct system. When no heating loads are present, follow the same procedure for heating ducts and hot deck. It should be noted that operating a dual-duct system as a single-duct system reduces air flow, resulting in increased energy savings through lowered fan speed requirements.

Single Zone System—A zone is an area or group of areas in a building which experiences similar amounts of heat gain and heat loss. A single zone system is one which provides heating and cooling to one zone controlled by the zone thermostat. The unit may be installed within or remote from the space it serves, either with or without air distribution ductwork.

1. In some systems air volume may be reduced to minimum required therefore reducing fan power input requirements. Fan brake horsepower varies directly with the cube of air volume. Thus, for example, a 10% reduction in air volume will permit a reduction in fan power input by about 27% of original. This modification will limit the degree to which the zone serviced can be heated or cooled as compared to current capabilities.
2. Raise supply air temperatures during the cooling season and reduce them during the heating season. This procedure

reduces the amount of heating and cooling which a system must provide, but, as with air volume reduction, limits heating and cooling capabilities.

3. Use the cooling coil for both heating and cooling by modifying the piping. This will enable removal of the heating coil, which provides energy savings in two ways. First, air flow resistance of the entire system is reduced so that air volume requirements can be met by lowered fan speeds. Second, system heat losses are reduced because surface area of cooling coils is much larger than that of heating coils, thus enabling lower water temperature requirements. Heating coil removal is not recommended if humidity control is critical in the zone serviced and alternative humidity control measures will not suffice.

Multizone System—A multizone system heats and cools several zones—each with different load requirements—from a single, central unit. A thermostat in each zone controls dampers at the unit which mix the hot and cold air to meet the varying load requirements of the zone involved. Steps which can be taken to improve energy efficiency of multizone systems include:

1. Reduce hot deck temperatures and increase cold deck temperatures. While this will lower energy consumption, it also will reduce the system's heating and cooling capabilities as compared to current capabilities.

2. Consider installing demand reset controls which will regulate hot and cold deck temperatures according to demand. When properly installed, and with all hot deck or cold deck dampers partially closed, the control will reduce hot and raise cold deck temperature progressively until one or more zone dampers is fully open.

3. Consider converting systems serving interior zones to variable volume. Conversion is performed by blocking off the hot deck, removing or disconnecting mixing dampers, and

adding low pressure variable volume terminals and pressure bypass.

Terminal Reheat System—The terminal reheat system essentially is a modification of a single-zone system which provides a high degree of temperature and humidity control. The central heating/cooling unit provides air at a given temperature to all zones served by the system. Secondary terminal heaters then reheat air to a temperature compatible with the load requirements of the specific space involved. Obviously, the high degree of control provided by this system requires an excessive amount of energy. Several methods for making the system more efficient include:

1. Reduce air volume of single zone units.
2. If close temperature and humidity control must be maintained for equipment purposes, lower water temperature and reduce flow to reheat coils. This still will permit control, but will limit the system's heating capabilities somewhat.
3. If close temperature and humidity control are not required, convert the system to variable volume by adding variable volume valves and eliminating terminal heaters.

THE ECONOMIZER CYCLE

The basic concept of the economizer cycle is to use outside air as the cooling source when it is cold enough. There are several parameters which should be evaluated in order to determine if an economizer cycle is justified. These include:

- Weather
- Building occupancy
- The zoning of the building
- The compatibility of the economizer with other systems
- The cost of the economizer

What Are the Costs of Using the Economizer Cycle?

Outside air cooling is accomplished usually at the expense of an additional return air fan, economizer control equipment, and an additional burden on the humidification equipment. Therefore, economizer cycles must be carefully evaluated based on the specific details of the application.

Using outside air to cool a building can result in lower mechanical refrigeration cost whenever outdoor air has a lower total heat content (enthalpy) than the return air. This can be accomplished by an "integrated economizer" or enthalpy control. See Figure 1-10 for a comparison of controls.

Dry Bulb Economizer

Operation of the "integrated economizer" can be made automatic by providing (1) dampers capable of providing 100% outdoor air, and (2) local controls that sequence the chilled water or DX (direct expansion) coil and dampers so that during economizer operation, on a rise in discharge (or space) temperature, the outdoor damper opens first; then on a further rise, the cooling coil is turned "on."

Economizer operation is activated by outside air temperature, say 72°F DB*. If outside air is below 72°F, the above described economizer sequence occurs. Above 72°F, outside air cooling is not economical, and the outdoor air damper closes to its minimum position to satisfy ventilation requirements only.

Enthalpy Control

If an economizer system is equipped with enthalpy control, savings will accrue due to a more accurate changeover point. The load on a cooling coil for an air handling system is a function of the *total heat* of air entering the coil. Total heat is a function of *two*

---
*This varies according to location.

Heating, Ventilating, Air Conditioning, and Building System Optimization 39

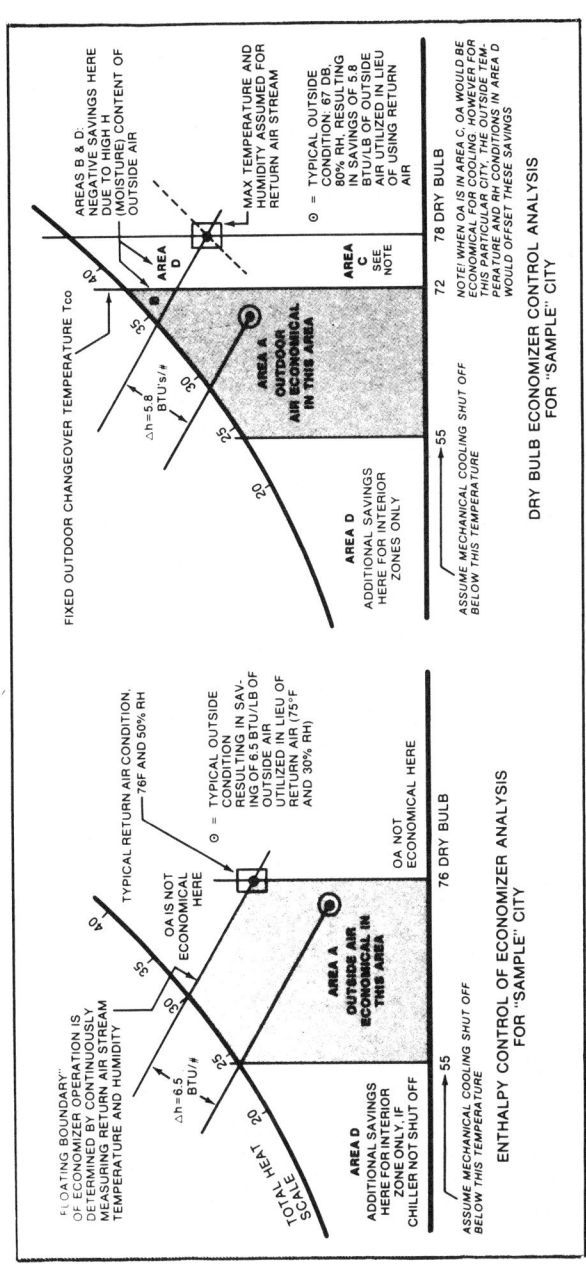

Figure 1-10. Comparative Economizer Controls

measurements, dry bulb (DB) and relative humidity (RH) or dew point (DP). The enthalpy control measures both conditions (DB and RH) in the return air duct and outdoors. It then computes which air source would impose the lowest load on the cooling system. If outside air is the smallest load, the controller enables the economizer cycle. (Dry Bulb Economizer control savings will be less than those shown for enthalpy control, except in dry climates.)

Enthalpy Control Savings Calculations

Savings are based on the assumption that the system previously had either (a) no 100% O.A. damper or (b) a fixed minimum outdoor air setting (whenever the fan operates) sufficient for ventilation purposes; it is also assumed that (c) minimum outdoor air has already been reduced, and the minimum damper opening will be at the *new value*.

Step 1. Determine minimum cfm of outdoor air to be used during occupied hours.
Step 2. Calculate annual savings.

$$A \frac{ft^3}{min.} \left(1 - \frac{B\%}{100}\right) \times K \frac{10^6 \text{ Btu}}{yr\ 1000\ cfm} \times \frac{\text{operating hrs/wk}}{50} \times J \frac{\$}{10^6 \text{ Btu}}$$

= $ SAVED PER YEAR *Formula (1-25)*

where
- A = air handling capacity $\left(\frac{ft^3}{min.}\right)$
- B = present ventilation air (%)
- J = cost of cooling $\left(\frac{\$}{10^6 \text{ Btu}}\right)$
- K = seasonal cooling savings $\left(\frac{10^6 \text{ Btu}}{yr\ 1000\ cfm}\right)$

Formula 1-25 is used to calculate the savings resulting from enthalpy control of outdoor air. The calculated savings generally will be greater than the savings resulting from a dry bulb economizer. To estimate dry bulb economizer savings, multiply the enthalpy savings by .93.

SECTION II
HEAT PUMPS

2

Industrial and Commercial Heat Pump Applications

T.C. Niess

INTRODUCTION

The large-scale application of heat pumps in the United States had its beginning with the first known installation of an air-to-air heat pump in Reading, Pennsylvania, in 1932. Since that time, the air-to-air heat pump market has gradually grown until it now represents a significant portion of the year-round air-conditioning installations in residential and small commercial projects. A major consumer attraction is the heat pump's ability to supply both cooling and heating at a relatively small installed cost addition over cooling-only systems and slightly more over forced-air heating systems, which are predominant in the United States.

No market for "heating only" heat pumps existed prior to 1973. Also, few large heat pumps of any type were installed, since the price of energy (electric, gas, and oil) was too low to encourage owner investment.

Then came the oil/gas crisis of 1973, spearheaded by the oil embargo and completely revolutionizing the energy pricing and availability structure. One major effect was the development of interest in waste heat recovery and heat pumping, which could conserve energy. At the same time, solar and geothermal resources were studied, and heat pumping emerged as a better way to use these alternative energy sources.

Since 1973, major heat pump developments in the United States have occurred and have been commercialized on a large scale. This chapter will cover heat pump applications in large commercial, institutional, and industrial projects principally using the nonreversible water-to-water closed vapor compression cycle and open-cycle mechanical vapor compression. Air-to-air, air-to-water, and water-to-air heat pump developments and their applications (largely residential) are beyond the scope of this discussion.

HEAT PUMP CYCLE DEVELOPMENT

Basic developments have largely centered on four cycle types[1]. Mechanical vapor recompression, as shown in Figure 2-1, is one cycle. The most common use is steam compression in industrial facilities, where owning and operating costs make it attractive to use low-pressure waste steam or flash hot condensate return to subatmospheric steam and compress it for reuse in the plant. Economics are often favorable when low compression ratios are involved.

A second cycle, as shown in Figure 2-2, using the addition of an evaporator, finds use in petrochemical applications, particularly for distillation columns with close proximity of bottom reboilers and overhead condensers. This can be an alternative to multiple-effect evaporators and evaporation of fluids having a low boiling-point elevation.

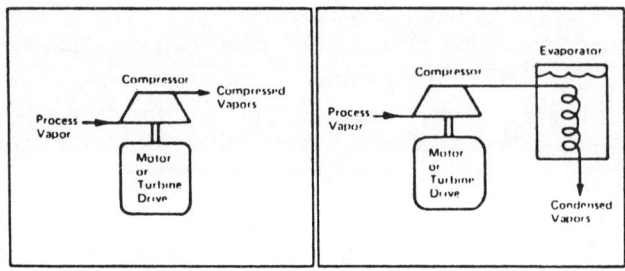

Figure 2-1. Open Vapor Cycle

Figure 2-2. Vapor Recompres- with Evaporator

Industrial and Commercial Heat Pump Applications   45

A third cycle, as shown in Figure 2-3, is the waste heat-driven Rankine cycle. This application is not yet widely used, largely due to high capital cost and some difficulties in matching heat source, heat sink, and compressed vapor needs.

The fourth cycle developed, as shown in Figure 2-4, is the closed vapor compression cycle. Using readily available fluorocarbon refrigerants as the heat pump working fluid, this cycle is commonly used because of its wide application opportunities.

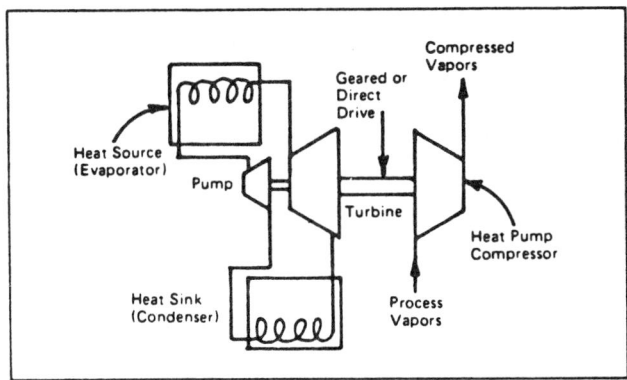

Figure 2-3. Waste Heat Driven Rankine Cycle

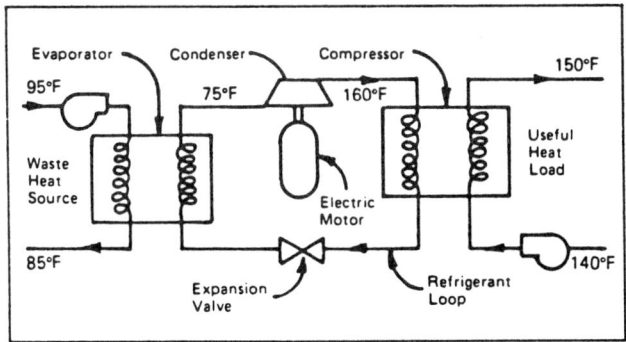

Figure 2-4. Closed Vapor Compression Cycle

CLOSED VAPOR COMPRESSION CYCLE

Typically, these heat pumps transfer energy from a waste heat fluid source, usually water, in a closed vapor compression cycle[2]. Low-temperature energy is recovered by evaporating the heat pump working fluid. These vapors are then compressed to a condensing temperature corresponding to pressure slightly higher than the delivery temperature. The process water stream then picks up this higher temperature energy in the heat pump's condenser, returning the condensed liquid working fluid through the pressure-reducing (expansion) valve to the evaporator. By increasing the heat of compression, more heat is delivered than is recovered from the waste heat source.

Commercially available in the United States, these closed-cycle heat pumps use reciprocating, screw, and centrifugal compressor-electric motor drives, usually hermetic. They are currently being applied up to 20-million-Btu/hr output capacities in a single unit. Due to the high cost of job-site labor and the requirement for high reliability, most units are factory assembled and tested prior to shipment.

The most commonly used heat pump working fluids in a reciprocating compressor are Refrigerant-22 up to 125°F, Refrigerant-500 up to 140°F, Refrigerant-12 up to 160°F, and Refrigerant-114 up to 220°F. In centrifugal machines with slightly lower upper temperatures, the most commonly used heat pump working fluids are Refrigerant-12 and Refrigerant-114. Refrigerant-113 is also used by some manufacturers.

The typical coefficients of performance (COP's) for large centrifugal compressor heat pumps are shown in Figure 2-5. These curves are based on leaving source water temperature from the evaporator and leaving hot water temperature from the heat pump condenser. These two temperatures, coupled with average delta temperatures found in the evaporator and condenser, determine the overall temperature lift. This, in turn, can be translated to power inputs and the COP by the formulas:

$$\text{COP} = F \left[\frac{T_u + 460}{T_u - T_s} \right] \qquad (2\text{-}1)$$

where:

$$COP = \frac{kW_t \text{ thermal output}}{kW_p \text{ power input}} \quad (2\text{-}2)$$

F   = heat pump performance factor

T_u = cycle condensing temperature °F

T_s = cycle evaporating temperature °F

460 = conversion to °R.

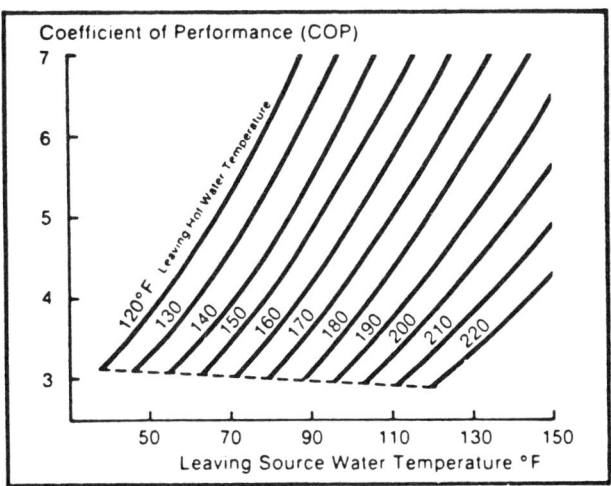

Figure 2-5. Typical Coefficients of Performance of Centrifugal Compressor Heat Pump

In applying large-capacity compressor heat pumps, the maximum temperature lift in a single unit is about 90°F. When a higher temperature lift is required, a two-stage cascade system, as shown in Figure 2-6, is normally used. This is done for several reasons—to take advantage of the better thermal characteristics of Refrigerant-12 at output temperatures of 140°F; to avoid operation at a vacuum at any point in the cycle; and to permit side loads to be handled at the intermediary loop temperature. This approach has proven to be more viable than multi-stage compression. Cascade systems have been installed in a number of projects, with significant success in performance and economics.

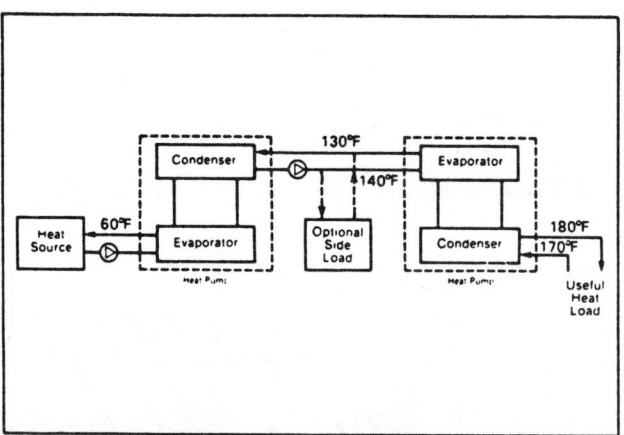

Figure 2-6. Cascade System

On smaller capacity installations, from about 14 kilowatts to about 100 kilowatts of power input, heat pumps using reciprocating compressors have generally been used.

WASTE HEAT SOURCES

The principle water heat source is water sent to a cooling tower which, in turn, rejects the heat energy to the atmosphere. Increasingly, this heat energy is being recovered and, through heat pumping, reused for space heating, service (domestic) water heating, and boiler feedwater and industrial process heating.

The waste heat can come from the condensers of refrigeration systems or air-conditioning systems for buildings or plants, low-level process heat from annealing and heat-treating furnaces, quench tanks, electric seam welders, air compressors, extruders, injection molders, pasteurizers, steam, and process condensers. Other sources that have been used include exhaust air streams and effluents from sewerage treatment plants and from manufacturing processes, such as those in textile, pulp, and paper mills.

More recently, where these sources are not available, there have been numerous installations using well water, groundwater, or low-temperature geothermal water from liquid-filled solar collectors and from earth-coupled heat pumps using buried pipes or downhole heat exchangers.

HEATING LOADS SERVED

Hot water from the heat pump condenser has been used for space heating, domestic water heating, makeup and reheat air heating, feedwater heating, parts washing and rinsing, degreasing systems, and general manufacturing processes that require process hot water.

Installations have been used in new construction and for retrofitting existing projects for increased energy efficiency. Percentage wise, they have been split about 50-50, with new construction seeming to be on the increase due largely to lower capital cost and increasing design attention to the "balanced heat recovery" concept by consulting engineers, architects, and owners.

TYPICAL COMMERCIAL INSTALLATIONS

In the United States, commercial installations include hotels, high-rise apartments and condominiums, and office buildings.

The principle need in hotels, apartments, and condominiums has been for the domestic hot water heating load. One example is the large Hershey Hotel in Philadelphia, Pennsylvania, which reclaimed heat from the air-conditioning system serving interior cooling loads in the lobby, restaurant, disco, meeting, and ball rooms. By using this waste heat, they—along with many other hotels—have heated most or all of the hot water needed for room, kitchen, and laundry use. In this case, the installation paid for itself in slightly over 1 year, operating at a 4.7 COP.

For intermittent hot water applications, such as domestic hot water, a storage tank is used. Typically, the storage tank is designed to serve one-half the peak hour demands, and the heat pump is sized to do the other half. Water is continuously circulated from the tank bottom, through the heat pump condenser and directly to the system or back to the tank, depending on the demand flow, as shown in Figure 2-7.

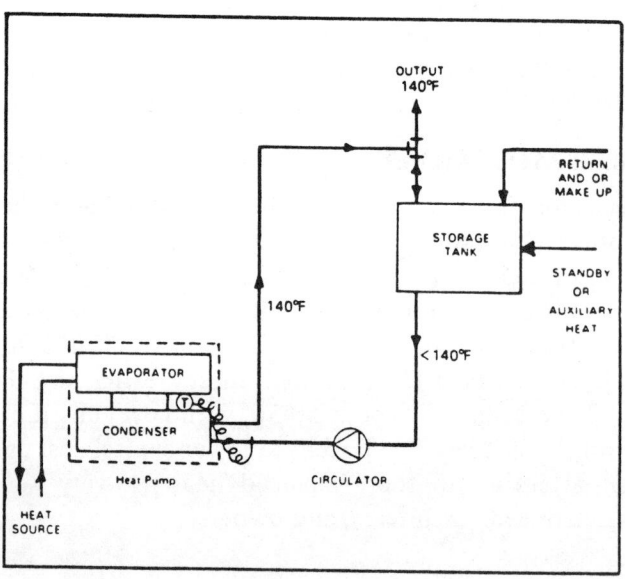

Figure 2-7. Domestic Hot Water Schematic

There have been numerous office building installations, typically in those that have a large internal cooling load year-round due to people, lights, computers, or any other heat-producing devices.

Such applications have ranged from a retrofit heat pump for domestic hot water and/or space heating to a multi-use installation typical of that used in several Bell Telephone Company buildings, as shown in Figure 2-8[3].

The waste heat from the chillers serving the year-round cooling load in the switch-gear areas is reclaimed and used to serve the perimeter heating load, reheat loads for humidity control and, through a heat exchanger, the building's domestic hot water needs. Three

Industrial and Commercial Heat Pump Applications 51

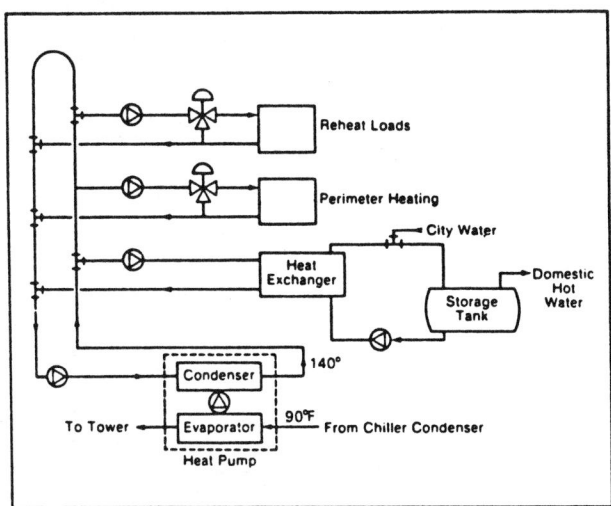

Figure 2-8. Typical Telephone Building Schematic

large centrifugal heat pumps—located on a mid-floor of the 20-story, 750,000-square-foot Indiana Bell Telephone Company headquarters—have split water head consdecers. Half of the hot water flow is piped to the upper floors, and the other half is pumped to the lower floors. This reduced the design water working pressure on the system that would otherwise have been required had the equipment been located in the sub-basement. The efficiency of this system is demonstrated by the fact that after 2 years of operation, a third identical unit was added to serve a second adjoining, newer building that was retrofitted for 150°F baseboard radiation from a constant-volume, all-air system.

Another rapidly emerging office building and hotel heat pump application uses a combination of multiple, unitary water-to-air heat pumps serving individual perimeter areas and cooling-only units serving interior zones, plus a water-to-water heat pump, as shown in Figure 2-9. The interconnecting loop is maintained between 60°F and 90°F. When the majority of the unitary heat pumps are in the heating cycle, the loop temperature gradually drops. When the loop temperature reaches 60°F, heat must be added. Conversely, when the units are predominantly cooling, the loop temperature rises. When it reaches 90°F, heat must be rejected. Within this range, heat is transferred back and forth via the loop for balanced heat recovery.

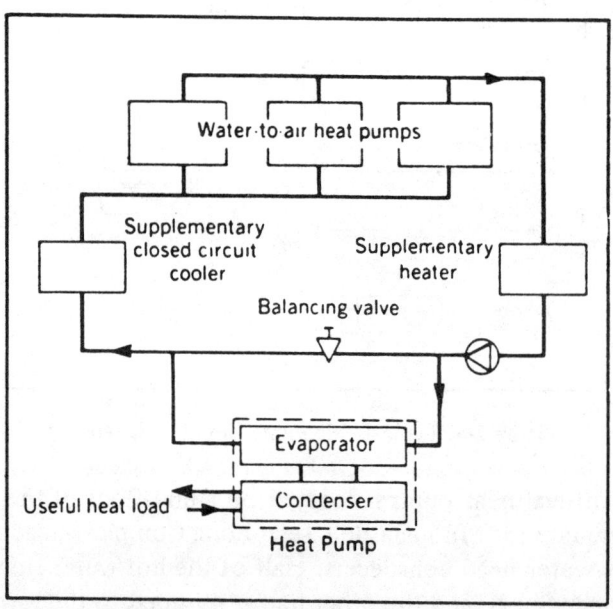

Figure 2-9. Combination System Schematic

Experience has shown that in many buildings, even in northern climates, the loop temperature in winter rises during the day, and heat is rejected. To effectively use this heat, a loop heat pump is added prior to the heat rejection device to generate hot water. Hotels use this to heat domestic hot water and/or makeup air.

Office buildings are storing this heat in a hot water tank, gradually heated to as high as 180°F during the day. In the 60,000-square-foot Philadelphia Life Insurance building built in 1980, a 15,000-gallon water tank is heated during the day whenever the loop temperature rises. When the automatic control system senses the loop temperature dropping, the heat pump is stopped, and the hot water from the tank is bled back into the loop to maintain its temperature above the minimum. This is shown schematically in Figure 2-10.

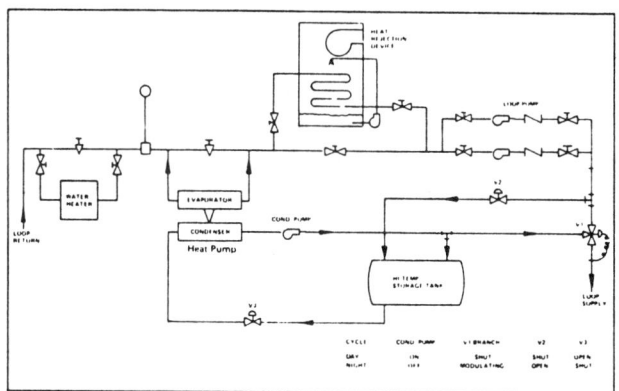

Figure 2-10. Combination System with High Temperature Storage

Only when the entire loop and tank system reaches 60°F is the water heating boiler turned on, and purchased energy used, to provide the building heat needed.

TYPICAL INSTITUTIONAL INSTALLATIONS

Hospitals have been a principle user of large-capacity heat pumps, primarily due to their need for hot water and the availability of waste heat. Hot water heating for hospital laundries consumes much of the thermal energy used. Waste heat is readily available from their chiller condenser and from the 75°F exhaust air. Heat pump reclaim from either of these sources is usually more than enough to provide the hot water needs. In some cases, both service water and space heating can be economically provided.

For example, Crozer Medical Center in Chester, Pennsylvania, uses 20,000 gallons/day of hot water in its laundry, which processes 5 million pounds of linens a year[4]. Incoming city water is first preheated to 75°F using a heat exchanger, supplied with 90°F water from a chiller condenser. This same waste heat source also serves the heat pump. During the laundry cycle, a 10,000-gallon tank is emptied twice a day. After refilling with the preheated 70°F city water, the heat pump gradually heats the tank to 150°F final temperature.

The mean COP for tank recovery over about 7 hours was 4.15, and the payback on the heat recovery investment was about 1½ years.

Universities, schools, laboratories, and museums have also used heat pumping effectively to reduce energy costs for space heating and domestic hot water heating. One example is Johns Hopkins University, which installed four large heat pumps over 6 years ago. The first installation of a centrifugal heat pump in August 1978 heats two buildings, totalling 160,000 square feet, and provides domestic hot water to five similar buildings, including two cafeterias[5]. Supply water temperature is reset based on outdoor temperature, from 120°F in warm weather (service water heating only) up to 160°F in very cold weather. During the first year of operation, fuel oil consumption was reduced by 40,000 gallons. Subsequent installation also proved equally attractive.

Another widespread heat pump application is providing space and service water heating in sewage treatment plants[6]. Prior to 1973, these plants used the gas byproduct from the digesters to provide heating. Designs today largely use this gas to fuel engine-driven pumps and generators. To meet these space heating needs, large heat pumps use the heat available in the plant's effluent and provide hot water to serve their heating needs.

Other institutional major-scale installations include buildings such as the large airport terminal complex at Orlando, Florida. The entire complex is both heated and cooled by a combination of water chillers and heat pumps. Standby heat has only been used when the cooling towers were shut down for routine maintenance or emergency repairs. Libraries, prisons, and government buildings are other institutional applications of closed-cycle heat pump heat recovery systems.

TYPICAL INDUSTRIAL INSTALLATIONS

Low-temperature heating within the output temperature range of current technology heat pumps has been used in a variety of industrial applications. Food processing industries use large quantities of hot water in the temperature ranges shown in Table 2-1.

Table 2-1
Food Processing Hot Water Uses

Process Use	Temperature °F
Cleaning and washdown	120 to 150
Bottle Washing	120 to 140
Pasteurization (beer, milk, etc.)	140 to 190
Sterilization	150 to 185
Cooking oil preheat	165 to 180

The waste heat source in food industries usually comes from condenser heat rejection of the refrigeration system[7,8]. Because this use of evaporative condensers is common in the United States, the installation usually includes adding a water-cooled refrigerant condenser piped in parallel with the evaporative condenser. The source heat needed is derived from this water and is piped to the heat pump. One alternative is a heat pump directly condensing the plant's refrigerant in the heat pump's evaporator, with the heat pump condenser delivering heated water. Principle installations to date have been in meat packing, dairy, brewery, poultry, and frozen food processing plants.

Many industrial plants use steam boilers having a relatively high boiler feedwater heating load. This occurs where steam is sent to the plant and is sparged into the process, becoming contaminated, or condensate is not returned for other reasons. A combination of heat exchanger/heat pump is then used to preheat incoming feedwater at 50°F to about 160°F, thus offloading the boiler from this low-temperature heating load.

Other industrial applications in which heat pumps have been commercially applied are parts washing, degreasing and rinsing in metal fabricating industries, plastic molding plants, electroplating and recovery operations, and glass fabricating plants.

In the past several years, a number of dehumidification heat pumps have been installed for drying furniture-grade lumber. The slow buildup of heat in the kiln results in reduced warping, splitting, and checking of the product, while reducing operating costs. Dehumidification heat pumps are also now being used as fish dryers and for indoor swimming pools.

ALTERNATIVE ENERGY SOURCES

A number of solar collector and groundwater/geothermal water well installations are currently in use in combination with large water-to-water heat pumps. In both cases, a sizeable reduction in the number of solar collectors or well water flow and/or temperature has been made possible by combining the alternative source with heat pumping.

Many of the large solar installations designed to produce hot water involved government financing or tax incentives, since the solar energy economics do not yet deliver an acceptable return on investment. One improvement in reducing the installed first costs has been coupling lower cost, liquid solar collectors with heat pumps.

As the solar heat collected can be well below the end-use temperature required, fewer and lower cost collectors can be used. The resultant first cost is then substantially lowered, with only a small penalty in operating cost.

One example of this concept is the solar heat pump installation at Mercy Hospital in Pittsburgh, Pennsylvania[9]. The solar system was originally intended to only meet the hospital's service hot water needs. The bids on this design came in at $980,000—which was well above available funding, even with the government grant.

The project was redesigned using a heat pump combined with fiberglass-covered polypropylene absorber collectors. With this system design, the bids were $504,000, and the project proceeded. One interesting and unanticipated benefit of the system was the ability to collect enough heat at night during the spring and fall to meet the lower nighttime hot water demand.

Groundwater and Geothermal Water Sources

The application of water-to-water and water-to-air heat pumps to extract energy from groundwater and surface water and from well water is becoming increasingly popular in the United States.[10, 11] Since groundwater and well water have temperatures ranging from 40°F to 80°F throughout most of the continental area, heat pumps are being used effectively for both heating and combination heating and cooling purposes.

In smaller capacities, the combination systems use refrigerant-reversing valves. In larger sizes, the water flow is reversed.

A unique application of a water-to-water heat pump is the Ephrata, Washington, geothermal energy project. The municipal water system is supplied from 84°F wells. A centrifugal heat pump extracts 13°F from the water, which is then returned to the city main, and supplies up to 130°F hot water to heat the municipal office buildings. In addition to the economic savings, the heat pump lowers the water temperature, which is too warm for domestic consumption.

In Europe, district heat systems are using heat pumps to reclaim heat from well and lake water or wastewater treatment plant effluent, in sizes up to 400-million-Btu/hr output.

OPEN-CYCLE (MECHANICAL VAPOR RECOMPRESSION) HEAT PUMPS

Unlike closed-cycle heat pumps, the open-cycle applications are all in industrial processes, where large quantities of energy are rejected in the form of low-pressure waste steam or low-grade waste heat (Figures 2-1 and 2-2). Economic heat recovery is possible, despite the low temperature levels and the possible contamination of the steam, vapor, or fluid. Table 2-2 lists some of the many applications where open-cycle heat pumps can be used to recover waste heat, conserve energy, and reduce energy costs.

Table 2-2
Typical Open-Cycle Heat Pump Applications

- Pulp and Paper
    - Neutral sulphite semi-chemical (NSSC) pulping evaporators.
    - Kraft pulping digester blowoff steam recovery, black liquor evaporators.
    - Thermomechanical pulping refinement of steam recovery.
    - Paper-drying, flash steam recovery.
- Chemical and Pharmaceutical Manufacturing
    - Evaporation, concentration, and crystallization processes.

- Distillation and stripping.
- Drying and heating.
- Food Products
    - Contaminated steam recovery.
    - Evaporation, concentration, and crystallization processes.
    - Food drying.
    - Rendering.
    - Steam cooking.
- Petroleum Refining
    - Low-pressure waste steam recovery.
    - Distillation.
    - Stripping and extraction.
- Miscellaneous
    - Solvent recovery.
    - Waste concentration.
    - Excess steam recompression.
    - Desalination.
    - Hot water flashing.
    - Boiler feedwater preheating.

Through the open cycle, waste heat can be converted to low- or subatmospheric-pressure steam, then mechanically compressed to a useful pressure, or the steam or vapor can be directly compressed to the needed process level.

OPEN-CYCLE COMPRESSORS

Open-cycle heat pump developments have primarily involved improvements in the design of screw, mixed-flow and centrifugal steam, and vapor compressors. The cycles are often referred to as mechanical vapor compression (MVC) or recompression (MVR) cycles. The selection of the compressor type is largely a function of the compression ratio (CR) required and the actual cubic feet per minute (ACFM) of inlet steam or vapor flow.

The compression ratio is defined as:

$$CR = \frac{\text{outlet psia}}{\text{inlet psia}} = \frac{\text{outlet psig} + 14.7}{\text{inlet psig} + 14.7} \qquad (2\text{-}3)$$

And the actual flow in cubic feet per minute is defined as:

$$ACFM = \frac{\text{lb/hr steam flow} \times \text{cu ft/lb at inlet pressure}}{60 \text{ min/hr}} \qquad (2\text{-}4)$$

In the open-cycle compression process, several different thermodynamic paths may be followed:[12]

- Direct compression of two-phase inlet mixture to desired conditions.
- Direct compression of dry steam to desired pressure, followed by liquid injection to desired temperature.
- Multistage compression of initially dry steam with interstage liquid injection.

The compressor COP, when compression takes place along these three paths, is shown in Figure 2-11.

The first method follows the most efficient compression path because the vapor is continuously cooled by evaporation of the entrained liquid droplets. This is practical with a screw compressor. Paths 2 and 3 are generally used with single-stage and multistage centrifugal compressors. To insure that the steam is dry at the compressor inlet, about 10 percent of the superheated outlet vapor is recycled back to the inlet. This represents an additional equivalent percentage reduction in COP over the values shown in the figure.

In general, the positive screw compressor is used for up to five compression ratios and in sizes from 1,200 to 20,000 ACFM in a single unit. For steam compression, it has the advantage of being able to use direct water injection in the compressor itself for steam desuperheating.

Mixed flow and conventional centrifugal compressors are available in single-stage design. Compression ratios are usually limited to two per stage, with desuperheating done externally to each state.

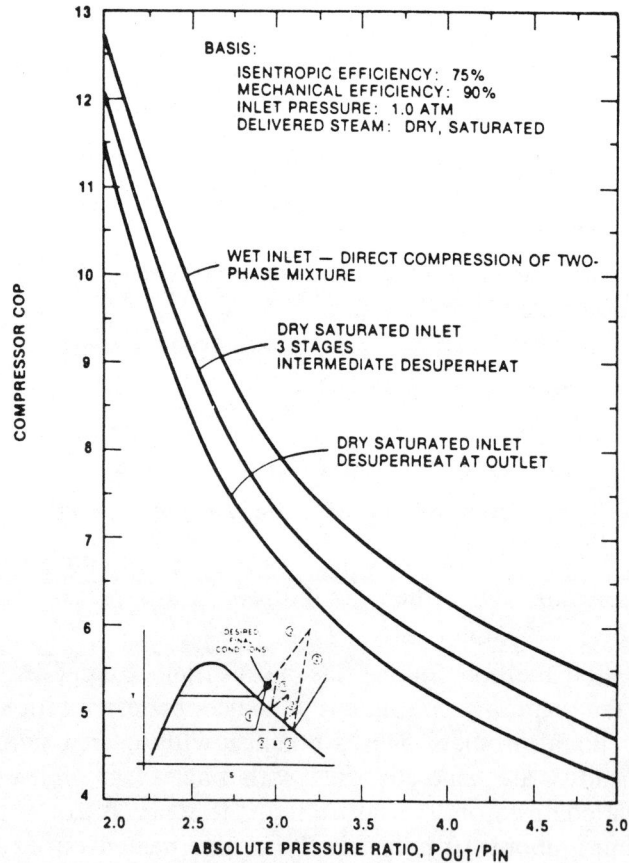

Figure 2-11. Compressor Coefficient-of-Performance at Various Compression Paths

Sizes start at about 2,000 ACFM, and range up to 75,000 ACFM for mixed flow and to over 200,000 ACFM for centrifugal machinery.

For rough estimating purposes, the compressor horsepower requirements can be approximated by the formula:

$$HP = \frac{15 \, (CR)(lb/hr \text{ steam flow})}{1,000} \qquad (2\text{-}5)$$

More precise figures can be obtained by actual calculation of the adiabatic head at design conditions and by using compressor and driver efficiencies obtained from the manufacturer.

The decision to use open-cycle versus closed-cycle heat pumps is usually a function of the process involved and the inlet conditions. Compressor costs are approximately linear to the ACFM. From the previously cited formula for ACFM, it can be seen that the ACFM will vary directly with the inlet cubic feet per pound.[4] Below about 160°F, the cubic feet per pound starts rising very rapidly by making the open cycle increasingly costly. Therefore, open-cycle steam compression is usually only considered above 150°F inlet conditions. Closed-cycle heat pump evaporation is economical for concentration of solutions that require low processing temperatures, down to 60°F.

EVAPORATOR TECHNOLOGY

Improvements have been made in the design and application o of evaporators with the open-cycle heat pump.[13]

Evaporation is a technology suitable for a broad range of processing applications in the food, pulp and paper, and chemical industries:

- Concentration of products.
- Chemical recovery.
- Concentration for incineration and heat recovery.
- Waste concentration.
- Crystallization.

Evaporators come in many configurations. The workhorse is the falling film evaporator, available in energy-conserving designs, incorporating multiple-effect configurations as well as mechanical vapor compression. Forced circulation evaporators and distillation columns complete the array of evaporation technology.

FALLING FILM EVAPORATORS

Falling film evaporators provide high heat transfer coefficients and maximize available temperature differences to reduce heat transfer surface requirements, allowing compact design. These evaporators may be made with the heat transfer surfaces either horizontally or vertically oriented.

Rising fuel costs have had a major impact on the economics of evaporator design. The first conservation measure was to link two or more evaporators in a series, so that heat resources could be used more efficiently. These multiple-effect units have gone to ever-larger numbers of evaporator bodies, as fuel costs made larger investment costs more practical. Mechanical vapor compression, coupled with these evaporators, can eliminate or substantially reduce steam requirements.

VAPOR COMPRESSION EVAPORATION

Mechanical vapor compression is ideally suited to handle fluids with low boiling-point elevation and provide the lowest energy consumption of any evaporative process—typically about 40 Btu/lb of water evaporated. This process is similar to multiple-effect evaporation, except that vapor is compressed and returned to the same module instead of flowing to additional effects (Figure 2-12). The

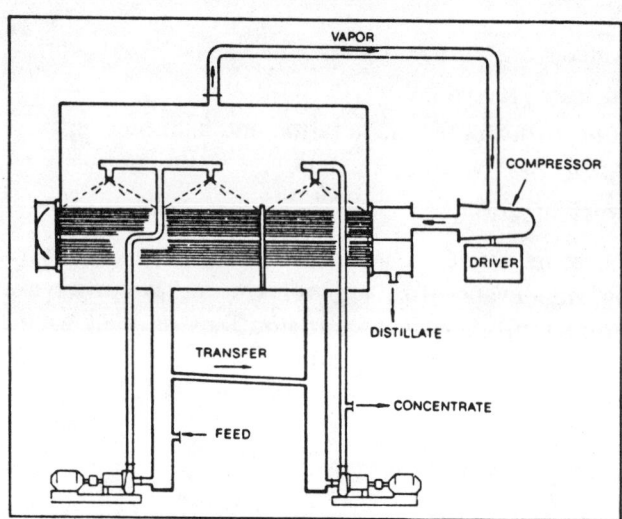

Figure 2-12. Low-Energy Vapor Compression System

vapor condenses on the inside of the heating tubes, providing the heat for vaporizing part of the recirculating fluid. Vapor compression evaporators are often added to increase the capacity and to reduce the energy consumption of existing systems.

DISTILLATION

Distillation is the separation of two or more components on the basis of the difference in their boiling points or volatility. This is accomplished by partial vaporization and subsequent condensation of the lower boiling-point components from those having a higher boiling point. Distillation systems are used for separation and enrichment of aqueous organic feed streams such as dilute ethanol or methanol, glycol, and solvents. Mechanical vapor compression is employed to reduce energy required for distillation by substituting mechanical energy for boiler steam to provide the heat of vaporization, as shown in Figure 2-13. As fossil fuel prices escalate, this technique will generate greater and greater operating cost savings, providing rapid capital cost payback.

Distillation equipment can also be integrated with evaporators to provide complete processing systems for such applications as agriculturally produced alcohols.

OPEN-CYCLE OPERATING ECONOMY

The operating economy of a multiple-effect evaporator system, compared to an open-cycle MVR heat pump system, depends on several factors.

- *Number of effects* — In general, about 0.85 pounds of water will be evaporated per pound of steam used. Therefore, the greater the number of effects, the better the economy.

- *Total Temperature difference* — The temperature difference between the supply steam and the condenser cooling water is the driving force for the multiple-effect evaporator. This difference must be divided between the effects. As each effect must have a reasonable temperature difference between

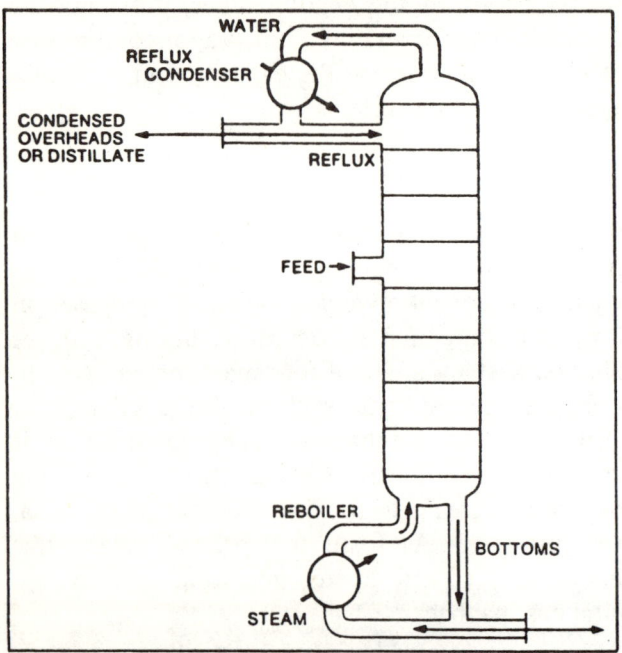

Figure 2-13a.
Conventional
Steam Distillation

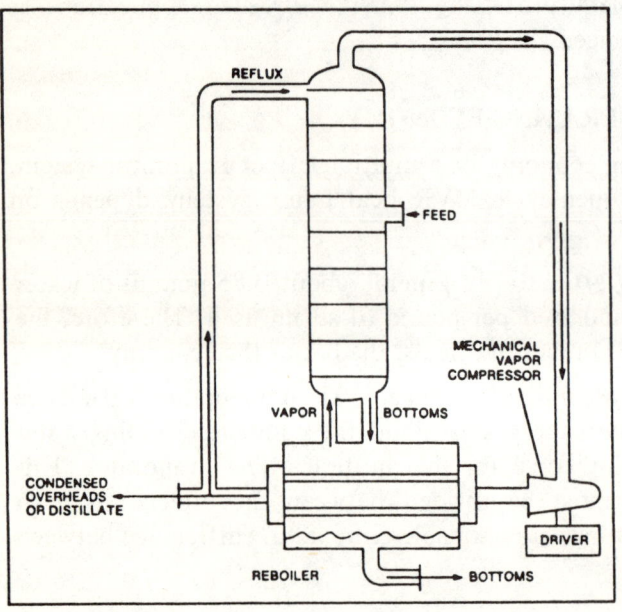

Figure 2-13b.
Vapor Compression
Distillation

vapor and product to cause heat transfer, there is a limit to the number of effects.
- *Boiling-point rise* — Pure water boils at 212°F at atmospheric pressure. When the water contains soluables, depending on their nature and the amount in solution, it will boil at a higher temperature—this difference being the boiling-point rise, or elevation.
- *Temperature difference across mechanical vapor compressor* — This is a direct function of the compression ratio. The lower the temperature rise (therefore, compression ratio) across the compressor, the better the operating economy.
- *Cost of energy* — This is the cost of steam (for the multiple-effect evaporator) compared to the cost of power (for the MVR heat pump).

TYPICAL ENERGY COST ANALYSIS

One example of such an analysis would be a pulp mill that wants to examine the cost of energy for concentrating a sulphite liquor from 10 to 60 percent by weight of total solids, comparing a five-effect evaporator to an MVR heat pump. At the boiler's net steam to PSIG action efficiency and current fossil fuel cost, the 25 psig steam costs $6.00 per million Btu and has a latent heat of 934 Btu/lb; electric power costs are $0.04 per kilowatt-hour.

FIVE-EFFECT EVAPORATOR

Typically, evaporation processes will operate at 0.80 to 0.90 pounds of water evaporated per effect per pound of steam used. For this analysis, 0.85 was used. The cost per 1,000 pounds of water evaporated is:

0.85 lbs water/lb steam • effect x 5 effects =
 4.25 lbs water evaporated/lb steam

$$\frac{934 \text{ Btu/lb steam}}{4.25 \text{ lbs. water/lb steam}} = 220 \text{ Btu/lb water evaporated}$$

$$\frac{\$6.00}{1 \text{ mill. Btu}} \times 1{,}000 \text{ lb water} \times 220 \text{ Btu/lb} = \$1.32/1{,}000 \text{ lbs}$$

This energy cost of $1.32 per 1,000 pounds of water evaporated will be compared to the cost of the process using a MVR heat pump.

MVR HEAT PUMP

Steam compressors will typically require 0.6 to 0.65 kilowatt input per degree F delta T per 1,000 lb/hr vapor flow for centrifugal design and 0.67 to 0.71 kW/F x 1,000 pph for screw compressors.

Considering the boiling-point rise of the sulphite liquor and a reasonable heat transfer temperature difference of a multiple-body spray film evaporator, the delta T saturated was determined to be 16.5°F.

The operating economy of the MVR heat pump is then:

0.65 kWh/°F · 1,000 lb x 16.5°F = 10.7 kWh/1,000 lb water

The energy cost of the MVR heat pump is then:

10.7 kWh/1,000 lb water x $0.04/kWh = $0.43/1,000 lb.

This represents an energy cost savings of:

$1.32 − $0.43 = $0.89/1,000 lb water evaporated.

The capital cost difference between these systems will typically result in a simple payback of under 2 years.

CONCLUSION

The increasing need in the United States to conserve energy and to reduce dependence on oil and gas has resulted in the use of a variety of heat recovery products. The use of heat pumps—which have been applied in many different industrial process heating, institutional, and commercial facilities—is prominent. New heat pump products and applications are limited only by the imagination of the engineer.

In the near term, it is expected that heat pumps will be applied to serve multiple-building, central heating and district heating projects. Source heat from low-temperature geothermal resources, and possibly from electric power plant waste heat, is the most likely to be used. Vapor recompression cycle developments will increase the applications for production and use of steam and in various distillation and evaporation processes.

Heat pumps are expected to continue their market growth, as energy prices increase faster than general inflation and economics dictate their consideration as an alternative to fossil fuels.

REFERENCES

[1] J.S. Gilbert, "Industrial Heat Pumps," *Plant Engineering*, November 23, 1983.

[2] "Applied Heat Pump Systems," *ASHRAE Handbook*, Systems Volume.

[3] P. Norelli, "Industrial Heat Pumps," *Plant Engineering*, August 21, and September 18, 1980.

[4] "Hospital Heat Recovery System Saves a Bundle," *Commercial Remodeling*, August 1981.

[5] A. Stucki, "Heat Pump Cuts Oil Use," *Contractors Electrical Equipment*, November 1979.

[6] R.C. Niess, "Effluents and Energy Economics," *Water/Engineering and Management*, August 1981.

[7] J. Forwalter, "Waste Heat from Refrigeration System Provides Energy for Hot Water," *Food Processing*, December 1979.

[8] D.Woods and R.F. Ellis, "Waste Heat Recovery System," *Food Processing*, August 1983.

[9] A. Weinstein, *et al.* "Solar Assisted Domestic Hot Water Heat Pump System," proceedings of the American Section of the International Solar Energy Society, May 1981.

[10] G.W. Lund, "Low-Temperature Geothermal Development," *Geothermal Resources Council Bulletin*, March 1984.

[11] R.C. Niess, "Geothermal Heat Pump Systems," *Geothermal Resources Council Bulletin*, December 1982.

[12] F.E. Becker and A.L. Zakak, "Mechanical Vapor Recompression for Waste Energy Recovery," proceedings of the National Petroleum Refiners Association Annual Meeting, March 1985.

[13] Aqua-Chem, Inc. bulletin WT-447-R1.

3

The Earth-Coupled Heat Pump As a Demand-Side Planning Tool

D.P. Mehta, J.G. Moroz, D.C. Zietlow

INTRODUCTION

Future demand for electricity has been treated traditionally as a predetermined quantity by utility planners[1]. Their job has been to estimate that quantity, then to plan supply accordingly. But the energy disruptions of the 1970s put a crimp on this familiar process. Predictable demand and flexible low-cost supply, the prerequisites of traditional planning, are no longer available for developing proper plans.

Demand-side planning carefully pin points utility action that can change customer demand in mutually beneficial ways. It is not just for reducing loads or just for building loads. It involves both—and all the load redistribution options in between.

For utilities with strong load growth, curtailing demand can defer the need for costly new construction. For those with ample reserve margins, building load can improve the return on investments already made. Even those utilities with a good overall match between capacity and demand can cut operating costs by redistributing demand more evenly throughout the hours of the day or the days of the year.

Although the possibilities for changing load shapes are infinite, five general types of change illustrate the range. The first three are classic load management techniques for improving utility load curves by smoothing out the peaks and valleys of customer demand. Peak clipping reduces system peak loads, valley filling build off-peak loads, and load shifting moves demand from on-peak to off-peak periods.

Another possibility, strategic conservation, reduces total energy use without necessarily reducing peak demand. In choosing this objective, the utility planner takes into account the conservation actions that would occur naturally and then evaluates the cost-effectiveness of utility programs to stimulate or accelerate those actions. The fifth possibility for changing a utility's load profile is strategic load growth, which means an increase in beneficial sales.

In the last two cases the strategic aspect is selectivity. Such potential benefits need to be quantified by evaluating the performance of electricity consuming products. For this purpose, the Central Illinois Light Company (CILCO) has initiated research in collaboration with Bradley University. Currently, two electricity consuming products viz: Earth Coupled Heat Pumps and Heat Pump Water Heaters are being evaluated experimentally with an objective of determining their wide-use impact on load shapes and to quantify the potential benefits both to utility and to customers from such load shape changes. The purpose of this chapter is to report the results obtained so far from the Earth Coupled Heat Pump (ECHP) project.

IMPACTS OF ECHP ON LOAD SHAPES

The ECHP is expected to be accepted widely in central Illinois by the customers over the air-to-air heat pump (AAHP) because the energy efficiency of AAHP is limited by the local outdoor temperatures in winter. Such an acceptance of ECHP is expected to have impacts on peak clipping, valley filling, load shifting, and on load growth. For example, if electric space heating, AAHP, and central air-conditioning are replaced by ECHP on a large scale, it will result in seasonal peak clipping. Similarly, if gas-fired furnaces are replaced by ECHP, valley filling and load growth will occur. The peak clipping and valley filling can be combined for load shifting. For the utility, peak clipping will improve demand control and valley filling will improve load factor. The purpose of the research reported in this chapter was to quantify the improvements in demand control and in load factor.

EXPERIMENTAL FACILITY

An Earth Coupled Heat Pump (see Figure 3-1) was instrumented to obtain experimental data for quantifying its impact on the load shapes and to determine the energy cost savings to the home owner. The details of the sensors installed are described in Table 3-1.

The data acquisition system consisted of a microprocessor based data logger, two microcomputers, a cassette recorder, two intelligent modems, telephone lines, and a printer. The components at the data acquisition-storage station were connected in a Hewlet-Packard Interface Loop (HP-IL) as shown in Figure 3-2. The datalogger in the loop was also interfaced to an RS-232 modem for telecommunication of data to the remote receiving station using the telephone lines. The data were retrieved and analyzed at the remote station using an IBM-PC. A software package "DACC" was developed for initializing and for directing the system to measure, store, transmit, and to analyze several state variables.

DATA COLLECTION

Data were recorded in two files, viz: daily file and run file. The indoor temperatures, outdoor temperatures, and earth temperatures at various depths were recorded in the daily file. The data for the 'ON' cycles of the heat pump were recorded in the run file and for each cycle they included: inlet $CaCl_2$ temperature, outlet $CaCl_2$ temperature, kWh of pump, kWh of compressor, kWh of supply air fan, kWh of duct heaters, supply air temperature, and return air temperature.

DATA ANALYSIS

Two computer programs were developed for data analysis. These programs calculate the daily energy savings, and the daily energy transfers to the supply air as follows:

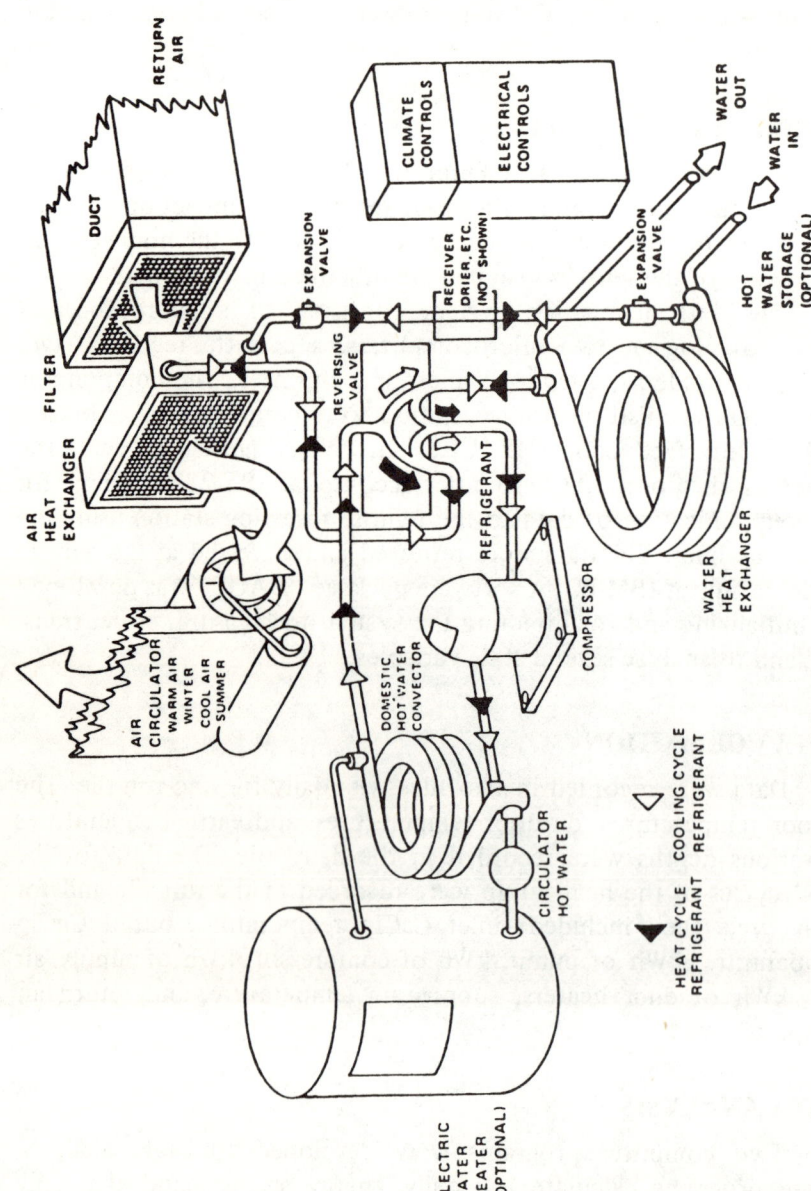

Figure 3-1. Schematic of the ECHP

Table 3-1. Details of Sensors

Variables to be Measured for	Description of Variables	Type of Transducer	Location of Transducer	No. of Points	Comments
I. Heat Load Calculations	Outside air temp.	T.C.	Outside of basement window	1	
	Inside air temp.	T.C.	Basement to 1st floor staircase	1	
	Air speed	None			Data taken from other sources
II. Earth Ground Storage Capacity	Earth ground temp. at various depths	T.C.	25 feet deep 40 feet deep 80 feet deep 160 feet deep	4	
III. Ground to H.P. Loop	Inlet water temp.	T.C.	Inlet water pipe	1	
	Outlet water temp.	T.C.	Outlet water pipe	1	
	Water flow rate	None			Assumed constant
	kWh's of pump	Opt. Iso.	240 to pump motor	1	kWh assumed constant Collect on/off status
IV. Heat pump Water to Freon Loop	Suction temp.	None			⎫ ⎬ For a later date ⎭
	Discharge Temp.	None			
	Flow rate of freon	None			
	kWh's to comp.	kWh - MA	240 to heat pump	1	
V. Freon to Air Exchange	Inlet air temp.	T.C.	Return air duct	1	
	Exit Air temp.	T.C.	Exit air duct	1	
	kWh of fan	Opt. Iso.	240 to fan	1	
VI. System Auxiliary	kWh's of the resistance heat	Opt. Iso.	240 to each element	4	kWh's assumed constant On/off status collected
TOTAL				17	

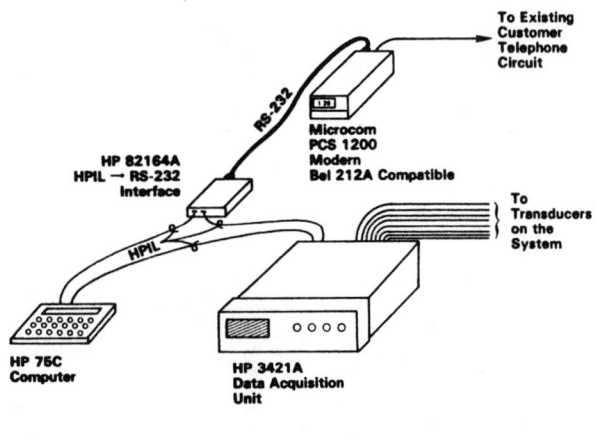

**Data Acquisition System
Hardware and Interconnections**

Figure 3-2.

A. **Energy Savings**

Daily energy savings are of great interest to the home owner and are best understood when translated into the dollar amount saved. These savings are reported in Table 3-2 and have been calculated using the data dollected. Savings for each run period of the heat pump were calculated and added over a period of 24 hours to get daily energy savings. Detailed procedures used for the calculations are shown in Appendix A.

B. **Energy Transfer**

Calculating daily energy transfer to the supply air is essential for evaluating the heat pump performance. Two performance parameters viz: coefficient of Performance (COP), and Heating Seasonal Utilization Efficiency (SUE) have been calculated and reported in Table 3-2. These performance parameters and their usefulness are described below.

Information on daily energy transfer to the supply air was also used for calculating the daily load factors for other heating systems like electric resistance heating for comparison. Detailed procedures for calculating the daily energy transfer to the supply air are shown in Appendix B.

C. **Electrical Demand Reduction**

The electrical demand reduction has been calculated by using the output of the computer program on the rate of energy transferred from the earth to the evaporator of the heat pump during the daily run file of minimum outdoor air temperature and the kW of the pump (K_p).

D. **Coefficient of Performance**

Daily coefficients of performance for the earth coupled heat pump have been calculated by using the computer output on the rate of energy transferred at the condenser of the heat pump and the rate of electrical energy consumed by the heat pump.

This parameter is useful to the home owner and to the utility company in comparing different makes and models of earth coupled heat pumps available.

E. **Heating Seasonal Utilization Efficiency (HSUE)**

HSUE is defined as the ratio of the daily energy consumed in electrical furnace to the daily energy input to the heat pump (2-4). The HSUE is averaged over the entire heating season and is assigned after extensive testing over a period of 5 years.

In this method of evaluating the performance of the heat pump, heat pump mode and electric resistance heating should be switched every 24 hours. However, no such switching was done during the period for which data have been analyzed in this report. Predicted values of HSUE have been calculated under the assumption that the heating loads were met using an electric furnace.

F. **Load Factors (LF)**

This is a very useful parameter for the utility company, for the utility LF is a measure of the ratio of revenues to the investment. Daily load factors have been calculated and reported in Table 3-2 while monthly LF are shown in Table 3-3. Detailed procedures for calculating these load factors are shown in Appendix C.

RESULTS

Table 3-2 tabulates the calculated values of kWh Saved, Dollars Saved, Electric Demand Reduction, COP, HSUE, Daily LF for ECHP and Daily LF for Electric Heating. Table 3-3 shows the calculations and the results for the Monthly LF for ECHP and for Electric Heating for the months November 1984 to March 1985.

CONCLUSIONS

An analysis of the results shown in Tables 3-2 and 3-3 shows that an ECHP is beneficial both to CILCO and to the Home Owner.

The major benefit to CILCO is an improvement in its Load Factors both on daily basis as well as on a monthly basis. On the average, daily load factors of ECHP are about 1½ times the load factors of an electric resistance heater. Similarly the monthly load factors using ECHP also showed improvements of the same order.

The Home Owner has also benefited a lot by having an ECHP installed and used. Energy savings for the period (91 days) for which the data have been analyzed so far were 9,200 kWh's which amount to financial savings of $368 @ $0.04 per kWh. Column 3 of Table 3-2 lists the daily dollar savings by the Home Owner by using ECHP compared to the use of Electric Heating.

Table 3-2. Daily Results

Date	kWh Saved	$ Saved	Electric Demand Reduction	C.O.P.	HSUE	L.F.H.P.	L.F.E.H.
11/28/84	87.42	3.49	8.15	3.11	2.75	0.49	0.42
11/29/84	75.20	3.00	8.09	3.18	2.71	0.40	0.36
11/30/84	83.35	3.33	10.85	3.16	2.74	0.43	0.41
12/01/84	91.31	3.65	7.99	3.16	2.68	0.49	0.44
12/03/84	133.02	5.32	8.09	3.20	2.67	0.72	0.63
12/04/84	139.18	5.56	7.91	3.19	2.59	0.76	0.64

(Continued)

Table 3-2. Daily Results *(Continued)*

Date	kWh Saved	$ Saved	Electric Demand Reduction	C.O.P.	HSUE	L.F.H.P.	L.F.E.H.
12/05/84	130.21	5.20	7.94	3.17	2.58	0.71	0.60
12/06/84	176.20	7.04	7.87	3.24	2.58	0.95	0.79
12/07/84	140.51	5.62	7.62	3.19	2.61	0.77	0.65
12/08/84	89.21	3.56	7.81	3.16	2.61	0.49	0.42
12/09/84	79.16	3.16	7.83	3.18	2.67	0.42	0.38
12/10/84	70.05	2.80	6.96	2.93	2.67	0.36	0.33
12/11/84	79.94	3.19	7.63	3.17	2.65	0.43	0.38
12/12/84	63.66	2.54	8.07	3.17	2.72	0.33	0.31
12/13/84	89.88	3.59	8.06	3.19	2.68	0.47	0.43
12/14/84	90.20	3.60	7.80	3.19	2.67	0.47	0.43
12/15/84	73.37	2.93	7.93	3.18	2.68	0.39	0.35
12/16/84	42.67	1.70	8.03	3.19	2.72	0.22	0.20
12/17/84	75.06	3.00	8.04	3.19	2.75	0.40	0.37
12/18/84	113.41	4.53	7.90	3.19	2.68	0.61	0.54
12/19/84	99.49	3.97	7.75	3.13	2.65	0.52	0.48
12/21/84	33.04	1.32	7.79	3.09	2.60	0.17	0.16
12/23/84	86.28	3.45	8.47	2.94	2.54	0.15	0.11
12/24/84	133.21	5.32	7.73	3.08	2.53	0.74	0.63
12/25/84	158.22	6.32	8.09	3.08	2.46	0.90	0.73
12/26/84	97.70	3.90	6.38	2.81	2.49	0.63	0.53
12/27/84	61.27	2.45	6.57	2.76	2.53	0.39	0.34
12/28/84	18.83	0.75	6.49	2.76	2.48	0.12	0.10
12/29/84	45.72	1.82	6.27	2.80	2.56	0.29	0.25
12/30/84	79.07	3.16	6.48	2.77	2.51	0.50	0.43
12/31/84	91.91	3.67	6.03	2.92	2.50	0.54	0.47
01/01/85	111.53	4.46	7.77	2.89	2.47	0.69	0.57
01/02/85	167.63	6.70	5.91	3.08	2.37	0.92	0.75
01/04/85	142.96	5.71	6.12	2.95	2.43	0.87	0.70
01/05/85	111.87	4.47	7.61	3.06	2.44	0.64	0.51
01/06/85	105.36	4.21	7.53	3.02	2.46	0.62	0.50
01/07/85	107.19	4.28	7.40	3.04	2.47	0.60	0.51
01/08/85	122.97	4.91	7.53	3.06	2.42	0.71	0.56
01/09/85	126.78	5.07	7.30	3.01	2.42	0.75	0.60
01/10/85	127.16	5.08	7.44	3.00	2.42	0.73	0.60
01/11/85	146.03	5.84	7.98	3.09	2.49	0.81	0.68
01/12/85	173.44	6.93	7.00	3.14	2.43	0.95	0.77
01/13/85	146.70	5.86	6.92	2.99	2.49	0.86	0.70
01/14/85	150.81	6.03	7.91	3.08	2.46	0.85	0.70
01/15/85	165.36	6.61	7.35	3.09	2.43	0.93	0.74
01/16/85	147.79	5.91	7.38	3.05	2.44	0.85	0.69
01/17/85	128.99	5.15	7.29	3.04	2.46	0.56	0.61
01/18/85	137.09	5.48	7.88	3.08	2.52	0.77	0.65

(Continued)

Table 3-2. Daily Results *(Continued)*

Date	kWh Saved	$ Saved	Electric Demand Reduction	C.O.P.	HSUE	L.F.H.P.	L.F.E.H.
01/21/85	80.65	3.22	6.65	2.87	2.50	0.51	0.42
01/25/85	84.18	3.36	7.33	3.04	2.41	0.48	0.39
01/26/85	153.10	6.12	6.81	2.88	2.47	0.96	0.75
01/27/85	130.20	5.20	7.39	3.00	2.49	0.77	0.63
01/28/85	153.84	6.15	7.15	3.06	2.59	0.88	0.76
01/29/85	131.90	5.27	6.58	2.96	2.70	0.82	0.71
01/30/85	110.49	4.41	6.69	2.97	2.73	0.66	0.60
01/31/85	167.29	6.69	6.84	3.11	2.74	0.96	0.85
02/01/85	153.65	6.14	6.54	2.72	2.83	0.99	0.89
02/02/85	144.89	5.79	6.28	2.73	2.78	0.89	0.84
02/03/85	131.48	5.25	6.12	2.66	2.81	0.98	0.80
02/04/85	119.57	4.78	5.83	2.71	2.72	0.88	0.75
02/05/85	98.15	3.92	6.71	2.69	2.75	0.70	0.62
02/06/85	136.53	5.46	6.61	2.95	2.75	0.85	0.75
02/07/85	139.74	5.58	6.33	2.81	2.73	0.94	0.81
02/08/85	124.42	4.97	5.69	2.71	2.72	0.89	0.77
02/09/85	114.57	4.58	5.96	2.69	2.73	0.83	0.72
02/10/85	92.38	3.69	5.93	2.82	2.72	0.61	0.54
01/11/85	108.25	4.33	6.78	2.78	2.78	0.73	0.66
02/12/85	127.05	5.08	7.36	2.96	2.78	0.78	0.70
12/14/85	130.52	5.22	6.67	2.93	2.76	0.82	0.73
02/15/85	125.86	5.03	6.48	2.85	2.82	0.84	0.75
02/16/85	95.25	3.81	5.92	2.74	2.78	0.67	0.59
02/17/85	81.77	3.27	6.63	2.90	2.66	0.39	0.44
02/18/85	81.32	3.25	6.68	2.90	2.68	0.50	0.45
02/19/85	83.05	3.32	6.78	2.91	2.69	0.52	0.46
02/21/85	57.05	2.28	6.87	2.96	2.78	0.34	0.31
02/22/85	36.47	1.45	6.87	2.67	2.77	0.25	0.23
02/23/85	26.58	1.06	5.67	2.64	2.81	0.18	0.17
02/24/85	50.14	2.00	5.25	2.56	2.83	0.39	0.35
02/25/85	66.01	2.64	5.41	2.78	2.79	0.44	0.40
02/26/85	72.33	2.89	7.04	2.96	2.79	0.44	0.40
02/27/85	90.53	3.62	7.01	2.98	2.76	0.55	0.49
02/28/85	72.24	2.88	7.01	2.96	2.77	0.45	0.40
03/01/85	62.63	2.46	6.95	2.94	2.82	0.37	0.35
03/02/85	58.31	2.33	6.88	2.96	2.80	0.35	0.32
03/04/85	76.04	3.04	6.74	2.96	2.85	0.47	0.43
03/05/85	94.62	3.78	6.52	2.92	2.77	0.58	0.53
03/08/85	13.21	0.52	7.39	2.90	2.60	0.08	0.07
03/09/85	42.89	1.71	7.23	3.06	2.61	0.25	0.21
03/10/85	35.81	1.43	7.30	2.90	2.71	0.22	0.18
03/11/85	45.47	1.81	7.14	2.74	2.65	0.30	0.27
03/12/85	38.22	1.52	5.74	2.66	2.69	0.27	0.23

(End)

Table 3-3. Monthly Load Factors

Month	Number of Days for which Data Used	Coldest Day	Min. Temp. on the Coldest Day (°F)	Time on (hrs)	Monthly Load Factors For E.C.H.P.	Monthly Load Factors For E.H.
November	3	11/30/84	26.40	0.03389	0.43	0.20
December	28	12/06/84	2.80	2.01306	0.53	0.22
January	25	01/31/85	−4.60	0.89139	0.80	0.32
February	26	02/01/85	−3.00	23.96389	0.61	0.29
March	9	03/05/85	20.20	0.22778	0.32	0.14

It is recognized that the above conclusions are based on limited amount of data and in comparison to Electric Heating only. Future data collection and analysis on additional facilities involving Air-to-Air Heat Pump and Natural Gas Heating Systems are expected to improve the quality and to enlarge the scope of our conclusions which will equip CILCO with reliable tools for the future demand planning.

ACKNOWLDEGEMENTS

We express our sincere thanks to the Central Illinois Light Company for providing the financial support for carrying out this research project. Assistance from Mr. H. Al Basha, graduate student at Bradley University, in data reduction is appreciated.

Appendix A:

Details of Procedures used for Calculating Daily Energy Savings

Daily energy savings are calculated from the following relation:

$$Q_{HP} = \sum_{r=1}^{n} \rho \, W \, c_p \, \Delta T_r \, P_r$$

$$= 8.4 \sum_{r=1}^{n} W \, \Delta T_r$$

where
> n = The number of daily run files.

$$\Delta T_r = T_{ir} - T_{or}$$

> T_{ir} = Average inlet $CaCl_2$ temperature (°F) for the rth run file.
>
> T_{or} = Average outlet $CaCl_2$ temperature (°F) for the rth run file.
>
> W = Flow rate of $CaCl_2$ (10.5 gpm)
>
> P_r = "ON" period for the rth run file.

Daily energy savings

$$K_s = \frac{Q_{HP}}{3,413} - K_p$$

where $K_p = \sum_{r=1}^{n} K_{pr}$

and K_{pr} = kWh of pump for the rth run.

Daily Dollar Savings = K_s x $0.04.

Appendix B:
Details of Procedures used for Calculating Daily Energy Transfers to Supply Air

Daily energy transfers to the supply air for evaluating the heat pump performance and load factors were calculated from the relation.

$$Q_{air} = \sum_{r=1}^{n} 1.09 \times CFM \times \Delta T_{ar} \times P_r$$

where $\dot{Q}_{air}$ = Daily energy transfer to the supply air.

CFM = Cubic Feet/minute flow of air.

$\Delta T as = T_{as} = T_{ar}$

T_{as} = Average supply air temperature for the rth run.

T_{ar} = Average return air temperature for the rth run.

Pr = 'ON' period for the rth run file.

Appendix C:
Details of Procedures for Calculating Load Factors (LF)

Daily load factors for the ECHP No. 1 have been calculated using the relation:

Daily LF for ECHP

$$= \frac{\text{Daily Sum of } (K_c + K_F + K_p + K_a) \text{ [kWh/Day]}}{\frac{(K_c + K_F + K_p + ka) \text{ for min. O.A. Temp run file [kWh]}}{\text{Duration of that run file [Hours]}} \times 24 \text{ [Hrs/Day]}}$$

and
Daily LF for Electric Heating

$$= \frac{\text{Daily Total kWh delivered to supply air [kWh/Day]}}{\frac{\text{kWh delivered to supply air for min. OA Temp file}}{\text{kWh delivered to SA for min. O.A. Temp run file}}}$$
$$\text{kW rating of the electric furnace} \times 24 \text{ [Hrs./Day]}$$

and
Monthly L.F. for ECHP

$$= \frac{\text{Monthly Sum of } (K_c + K_F + K_p + Ka) \; [\text{kWh/Month}]}{\dfrac{(K_c + K_F + K_p + Ka) \text{ for min. O.A. Temp run file [kWh]}}{\text{Duration of that run file (hours)} \times 24 \, [\text{Hrs./Day}] \times [\text{Days/Month}]}}$$

Monthly LF for Electric Heating

$$= \frac{\text{Monthly Total kWh delivered to Supply Air [kWh/Month]}}{\dfrac{(K_c + K_F + K_p + Ka) \text{ for min. OA Temp run file [kWh]}}{\text{Duration of that run file [hours]} \times 24 \, [\text{Hrs./Day}] \times [\text{Days/Month}]}}$$

In the above relations, K_c, K_F, K_p and K_a are the kWh values for the compressor, fan, and pump auxiliary heaters respectively.

LIST OF NOMENCLATURE

Symbols
C — Specific Heat (Btu/lbm°F)
K — Kilowatt 1 hours (kWh)
Q — Heat Transfer Rate (Btu/hour-time)
T — Temperature (°F)
W — Flow Rate (gallons/minute)
ρ — Density

Subscripts
A — Auxiliary
AMB — Ambient
F — Fan
HP — Heap Pump
TC — Thermocouple

i — Inlet
O — Outlet
P — Pump, pressure
r — Return
s — Supply

References

[1] Wayne, Mary. Demand Planning in the 80s. *Electric Power Research Institute J.* December, 1984. p. 6.

[2] Didion, D. and George Kelley, New Testing and Rating Procedures for Seasonal Performance of Heat Pumps. *ASHRAE J.* 1981 (9): 40.

[3] Hughes, J.A.; J. Kasprzcki and L.R. Pool. Using the Heat Pump Annual Utilization Efficiency to Optimize Design and Conservation of Energy. *ASHRAE Trans.* 86 (1) 1980: 660

[4] Documentation and Analysis of Improvements in Efficiency and Performance of HVAC Equipment and Systems. Ray W. Herrick Labs. School of Mechanical Engineering, Purdue University. Oct. 23-25, 1978.

4

Case Study: Heat Pump Heat Recovery System At Tend-R-Fresh

P.A. Rowles

INTRODUCTION

The key to the success of any heat recovery system is the degree to which the system can match the supply of waste heat with the demand for thermal energy. This is even more critical when heat pumps are used, due to their high capital and operating costs. As a result, in designing cost effective heat pump systems, it is necessary to minimize the capacity of the heat pump as well as the resulting electrical energy costs to run them.

This chapter discusses a heat pump/heat recovery system which has been developed by Balanced Energy Systems Technology Inc. to recover low-grade waste heat from refrigeration systems, and use it to produce process hot water. The system is well suited to food and beverage processors who simultaneously require cooling and hot water.

The system which is discussed here is installed in a poultry processing plant, known as Tend-R-Fresh, in the small community of Petersburg, Ontario. Tend-R-Fresh is owned by Maple Leaf Mills Limited whose head office is in Toronto, Ontario. Maple Leaf Mills purchased the system with partial funding provided by The Canadian Electrical Association and Ontario Hydro as a grant for research and development. The heat recovery system was installed during 1984 and officially started in December of that year. Its operation is being continuously monitored and optimized by Balanced Energy Systems Technology Inc.

The results of this monitoring and optimization are presented in this chapter. Also presented in this chapter are a description of the system installed at Tend-R-Fresh, and recommendations for improvements in the design for future systems.

BACKGROUND

Refrigeration plays a significant role in many food and beverage processes. Its purpose is to remove energy from the product. It is used for process chilling, blast freezing, and cold or frozen storage. The energy which is removed from the product must be exhausted in some manner, usually to the outside atmosphere through evaporative condensers. Despite the widespread use of refrigeration, the full potential of recovering this waste heat has not been realized. To assist in understanding the nature of the problem, a brief description of refrigeration systems and their energy balance is presented here.

Figure 4-1 shows a schematic diagram of a basic refrigeration system. There are four major components—the compressor, condenser, expansion valve and evaporator.

Basically, the compressor takes low pressure refrigerant gas, compresses it into a high pressure gas, and circulates it around the refrigeration circuit. The conderser receives the high pressure gas and changes it into a high pressure liquid by cooling it with air or water. The expansion valve controls the flow of high pressure liquid refrigerant into the low pressure evaporator. Here, the high pressure liquid refrigerant is allowed to expand or boil into a gas as it is released through the expansion valve. In doing so, the refrigerant absorbs heat from its environment. The refrigerant then returns to the compressor to start the cycle again.

Under typical operating conditions, approximately 12,000 Btus (12660 kJ) per hour are absorbed in the evaporator for every ton of refrigeration capacity. The compressor adds about 3000 Btus (3165 kJ) per hour per ton of refrigeration, increasing the temperature and pressure of the refrigerant. The energy which is absorbed in the evaporator and added by the compressor must be rejected by the condenser to maintain a balanced system. The energy rejected by the conderser amounts to 15,000 Btuh (15,825 kJ) per ton of refrigeration. This represents a significant source of waste heat. In a 100-ton refrigeration system, the condenser can reject up to 1.5 MMBH (1.695 GJ). In typical refrigeration systems, this energy is available at temperatures up to 95°F (35°C). In addition to exhausting large quantities of heat, refrigeration systems also use large amounts of water for evaporative cooling in their condensers.

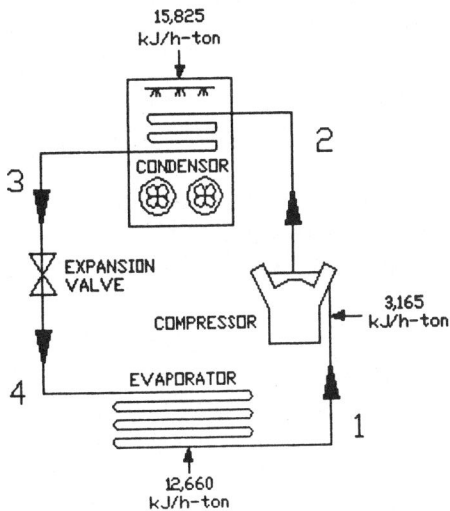

Figure 4-1. Basic Refrigeration System Energy Balance

THE HEAT RECOVERY SYSTEM

The heat recovery system installed at Maple Leaf Mills' Tend-R-Fresh plant is designed to:

1. Recover low-grade energy, in the range of 75-90°F (24-32°C), normally exhausted by refrigeration condensers.

2. Upgrade the recovered waste heat, using an electric high temperature heat pump, to supply process hot water in the range of 130-150°F (54-65°C).

3. Control the additional electrical load of the heat pump to take advantage of the low cost of off-peak power, using

thermal storage of process hot water and load control techniques.

The heat recovery system is illustrated schematically in Figure 4-2.

The existing evaporative condensers were replaced by shell and tube condensers on three icemaking systems operating in the plant. The combined capacity of the icemaking equipment represents about 90 tons of refrigeration. The waste heat recovered by the shell and tube condensers is estimated to be 1.35 MMBH (1.42 GJ/h), when all three icemakers are in operation. The icemakers were selected because of the flexibility in scheduling their operation. By scheduling the bulk of icemaking to off-peak hours, electrical costs for producing both ice and hot water could be reduced through avoidance of peak-demand charges.

The waste heat is recovered by glycol which is circulated through the shell and tube heat exchangers. Typically, the glycol will enter the shell and tube at 75°F (24°C), and be heated to 85°F (29°C). Using the shell and tube condensers eliminates both the need to run cooling fans, and the necessity of providing water for evaporative cooling. The shell and tubes also require minimal maintenance.

After recovering the waste heat from the condensers, which are located on the roof of the plant, the glycol is first used to preheat make-up water to the hot water system. This is done in a plate-type heat exchanger. Under typical operating conditions, the make-up water is heated from 46°F (8°C) to 75°F (24°C). The glycol is cooled from 85°F (29°C) to approximately 82°F (27°C). The glycol is then pumped through the evaporators of two heat pumps. Here the glycol is cooled, using the same refrigeration principles previously discussed, to a temperature of 75°F (24°C). From the evaporation of the heat pumps, the glycol is returned to the shell and tube condensers to recover more heat and repeat the cycle.

The energy absorbed from the glycol, in the evaporators of the heat pumps, is upgrated by the heat pump compressors. The condensers of the heat pump reject this upgraded energy by heating process hot water up to 160°F (70°C). The make-up water, which has been preheated in the plate heat exchanger, is mixed with hot

Case Study: Heat Pump Heat Recovery System at Tend-R-Fresh 89

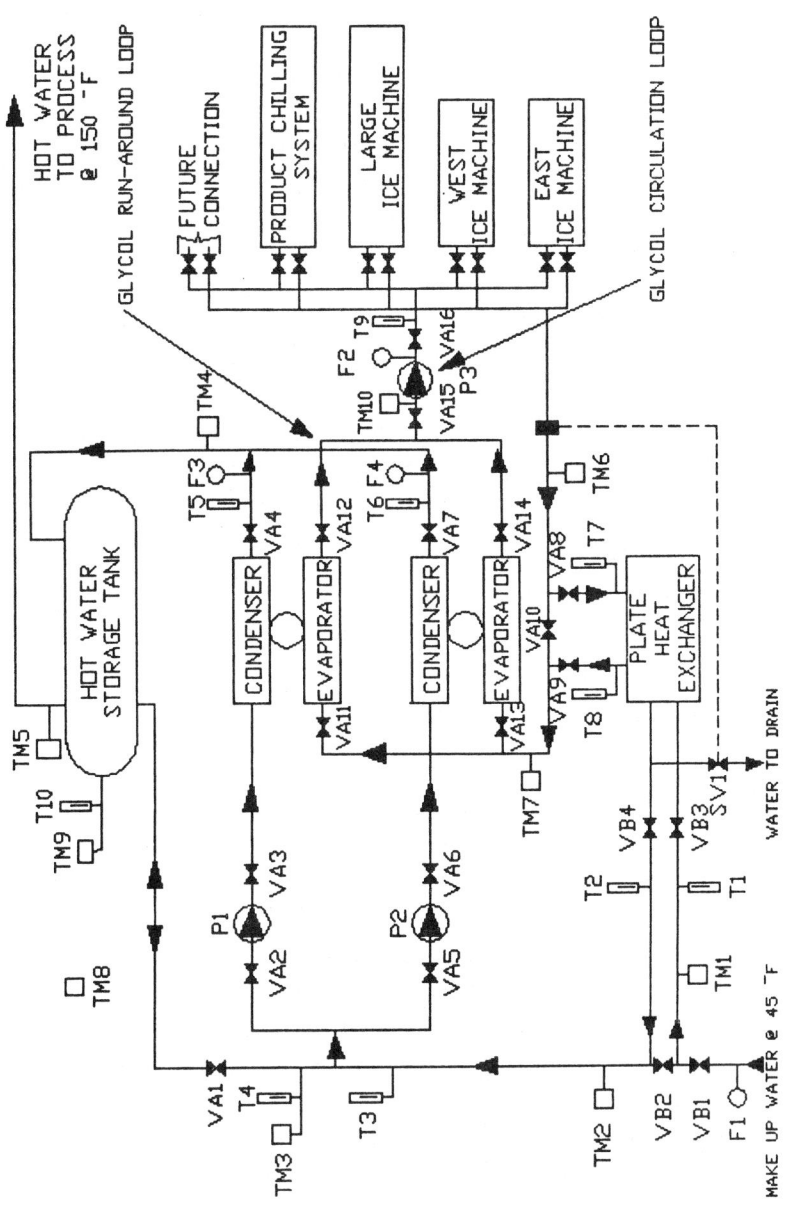

Figure 4-2. Heat Pump Heat Recovery System

water from the storage tank. This water is further heated by the heat pumps and delivered to the process. If there is no demand for hot water, the water is recirculated through the storage tank, thereby increasing its temperature. The storage tank ensures that hot water is available even when the heat pumps are not operating.

From this discussion it is clear that the heat pump system provides two functions:

1. Condenser cooling water for refrigeration system.

2. Hot water for processing.

Because of the resulting improvement in the operation of the refrigeration system, a decision was made to rely solely on the heat recovery system to provide condenser cooling. For this reason, a back-up was required for times when the heat pumps are not operating. This is achieved using the plate heat exchanger. When additional cooling is required, a valve is opened which allows more water to flow through the heat exchanger, cooling the glycol down to the desired temperature.

The hot water generated by the heat recovery system is used for several applications:

1. Preheated make-up water to the scalder.

2. Process hot water to the production line.

3. Hot water for plant cleaning.

The scalder operates at 140°F (60°C) and requires 10 gpm (45 lpm) of make-up water. Hot water, along the processing line, is required at 120°F (50°C) and cleaning water is 180°F (82°C). In most cases, heating was previously supplied by direct steam injection, or through a steam heated hot water tank. The heat recovery system currently supplies an average of 35,000 imperial gallons (160,000 litres) of hot water per day. The target is to supply up to 70,000 gallons per day.

LOAD MANAGEMENT SYSTEM

The operation of a heat pump in this waste recovery system will add 100 to 150 kW to the electrical load of the plant. If this load is

Case Study: Heat Pump Heat Recovery System at Tend-R-Fresh

operated during the peak demand interval, additional charges of approximately $423.00 (100 kW x $4.23) per month or, $5,076 annually will result. This can be avoided if load management techniques are used in the operation of this system. A load controller was installed to monitor the electrical demand of the plant continuously, and to disconnect the heat pumps during periods of peak power demand. The heat pumps are automatically reconnected when plant power levels fall below acceptable limits. When the heat pumps are not in operation, hot water will be drawn from the storage tank. Controlling the operation of the heat pumps in this manner has the effect of reducing the operating electrical costs of the heat pumps by reducing the impact of the demand charge.

SYSTEMS SPECIFICATIONS

Heating Capacity	2,327,000 Btuh (2.5 GJ/h)
Maximum Output Temperature	160°F (71°C)
Average Output Temperature	140°F (60°C)
Design Water Flow	35 Imp. gpm (2.65 l/s)
Cooling Capacity	1,985,000 Btuh (2.1 GJ/h)
Equivalent Refrigeration Cooling Capacity	132 Tons Refrigeration
Glycol Flow Rate	210 Imp. gpm (16 l/s)
Glycol Entering Temperature	85°F (29°C)
Glycol Leaving Temperature	75°F (24°C)
Electrical Load	
Heat Pumps	110.4 kW
Circulating Pumps	14.0 kW
Design Coefficient of Performance (C.O.P.)	
Heat Pumps	4.23
System	5.48

PERFORMANCE

The heat recovery system is being continuously monitored using a remote monitoring system developed by the B.E.S.T. Corporation. The micro processor based, remote monitoring system measures average temperatures, flows and kilowatt-hours over 15-minute intervals and stores this data in internal computer memory. The data is accumulated over a period of 24 hours, at which point the remote monitoring system dials the phone number of an Apple computer stationed in the B.E.S.T. office. When a connection between the remote unit and the central conputer is established, the data is transmitted by the remote unit. The central computer checks the data as it's received, and stores it on disk to be processed at a later time. The data is usually transmitted between 3 and 4 a.m.

The parameters which are being continuously monitored by the system are:

T1	Water temperature entering plate heat exchanger
T2	Water temperature leaving plate heat exchanger
T3	Water temperature entering heat pump
T4	Water temperature leaving heat pump
T5	Water temperature to process
T6	Water temperature in storage tank
T7	Ambient air temperature
T8	Glycol temperature entering plate heat exchanger
T9	Glycol temperature leaving plate heat exchanger/entering heat pump
T10	Glycol temperature leaving heat pump
F1	Water flow to process
KW1	Electrical energy use by heat recovery system
KW2	Total plant electrical energy use

This data is used to calculate the total thermal energy delivered to the process, as well as the electrical energy required to operate the system. From these numbers, an effective coefficient of performance (C.O.P.) or energy ratio can be determined. Figure 4-3 shows a typical graphic summary of the daily operation of the system. This graph shows hourly water flow to process, hourly thermal energy delivered to process, as well as hourly electrical energy consumption. On this particular day, the overall C.O.P. of the system was 5.3. That is 5.3 units of thermal energy were delivered for every unit of purchased energy. To date, the system has operated with an average C.O.P. of approximately 4.5, including weekends and holidays.

Case Study: Heat Pump Heat Recovery System at Tend-R-Fresh   93

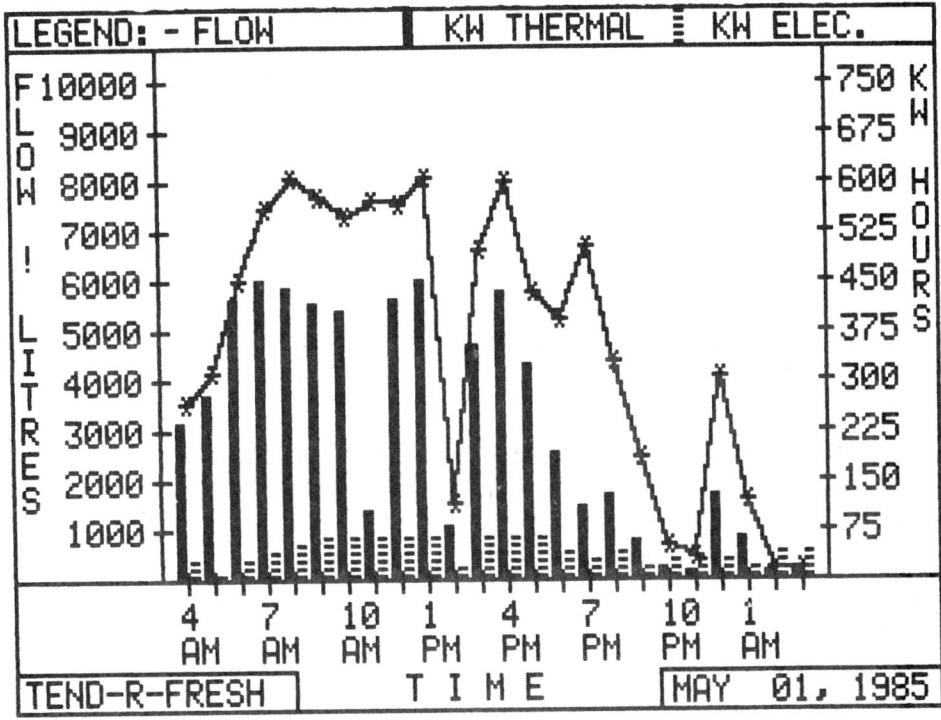

COST BENEFIT ANALYSIS   CONCLUSIONS

Figure 4-3. Summary of Daily Performance

The C.O.P. is being steadily improved through optimization procedures. Optimization to date has included:

- piping changes to improve water flow through system and to the process.
- adjusting temperature controls on heat pumps to ensure maximum free heat gain through plate heat exchanger and maximum C.O.P. in the heat pumps. This is achieved by raising the temperature level in the glycol circulating loop. There are trade-offs in adjusting the temperature of the glycol, as any overall increase in temperature raises the head pressure in the refrigeration system and reduces its performance.

- adjusting the sequencing on heat pump compressors to ensure that each compressor is running fully loaded, hence maximum efficiency when running.

Future changes to the system which are intended to improve its performance include:

- connection of additional heat sources from other refrigeration systems in the plant.
- load controlling heat pumps and refrigeration equipment to minimize electrical costs.

COST BENEFIT ANALYSIS

Energy Production
Average Daily Hot Water Usage	70,000 gallons (Imp)
Average Temperature	140°F
Average Make-up Water Temperature	45°F
Daily Energy Delivered to Process	66.5 MMBtu
Annual No. of Production Days	232
Annual Energy to Process	15428 MMBtu

Energy Savings
Displaced Fuel	Natural Gas
Price	
Heating System Efficiency	75%
Daily Reduction in Fuel Consumption	88.7 mcf
Annual Reduction in Fuel Consumption	20587.4 mcf
Annual Cost Saving	$112,152

Operating Cost
Daily Electrical Usage	3608 kWh
Annual Electrical Usage	837,104 kWh
Average C.O.P.	5.4
Average Electricity Price	$0.03/kWh
Annual Electrical Cost	$29.299

Economic Analysis
Net Cost Savings	$82,853
Capital Cost	$197,000
Payback	2.4 years

CONCLUSIONS

The results of this research and development project illustrate both the technical and economic viability of this type of heat recovery system. Through the recovery of waste heat from refrigeration systems and the subsequent upgrading of it using a high temperature heat pump, the system can deliver hot water to plant processes at temperatures up to 160°F (71°C). This is achieved with an overall coefficient of performance (C.O.P.) of 5.5. Under these conditions, the system provides a payback of 2 to 3 years. This does not include cost savings from improved performance of refrigeration equipment, reduced maintenance on evaporative condensers, and reduced water consumption by evaporative condensers.

RECOMMENDATIONS

As this project was a first-time demonstration of the technology, there were some problems and additional costs encountered which will likely not occur in subsequent installations. The physical layout of the existing refrigeration equipment made it necessary to use a glycol run-around loop, and three shell and tube heat exchangers. This increased the cost of the project, and reduced the efficiency of the system. A new system, which is currently being designed for another Maple Leaf Mills poultry plant in Brampton, Ontario, will eliminate some of the unnecessary and costly items that were included in the Tend-R-Fresh installation. A schematic flow diagram for this system is illustrated in Figure 4-4. The glycol run-around loop is eliminated in this system because the plant operates with a centralized refrigeration system, as opposed to the distributed one at Tend-R-Fresh. As a result, both the preheater, as well as the heat pump evaporator, can be connected directly in the hot gas refrigerant line. In doing so, the heat recovery system has access to all of the waste heat generated by the refrigeration system. The existing evaporative condensers act as back-up to the heat recovery system. The new system requires a heat pump which is designed specifically for this application. Heat pumps currently commercially available for industrial applications are either water- or air-based. The heat pump proposed for this application is designed to recover heat dir-

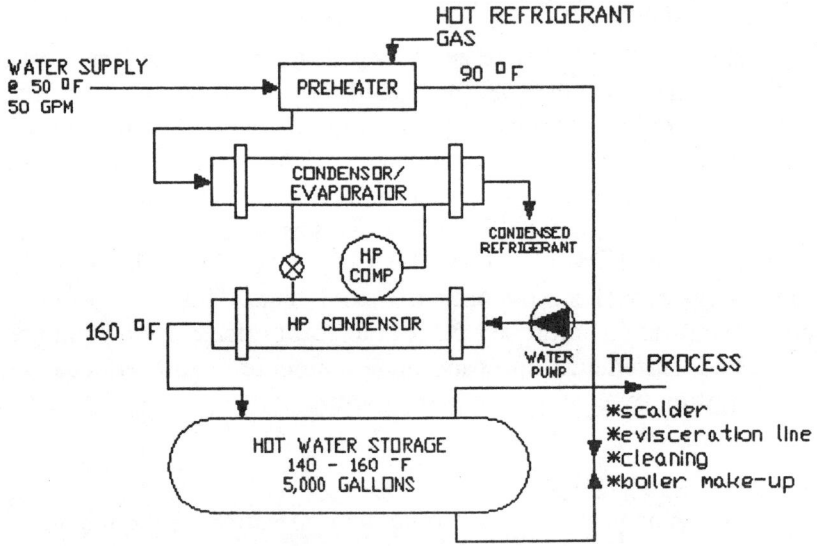

Figure 4-4. Schematic Diagram for Future Installations

ectly frrom the hot refrigerant gas of the existing refrigeration system. To provide a system with the same heating capacity as at Tend-R-Fresh, the estimated capital cost is $151,000. Eliminating the glycol run-around loop will increase the performance of the system to an estimated C.O.P. of 6.8. Under these conditions, the installation will pay for itself in 1.7 years.

This project has demonstrated the viability of this system in a poultry processing plant. The successful application of this system is not limited to poultry plants. The system can be adapted for use in other food and beverage applications. Virtually every plant that requires both refrigeration and hot water is a candidate. Specifically, meat processing, diary products, frozen foods and frozen vegetables are suitable. In view of the extensive use of refrigeration in food processing in North America, the energy saving potential of this system is enormous.

5

Case Study: Heat Pump Strategy At the Nevada Test Site

R.E. Davis

INTRODUCTION

The high desert location of the Nevada Test Site (NTS) provides both high and low temperature extremes during the course of a year. The remoteness of the NTS resulted in the selection of electricity as the principal energy source for environmental conditioning systems. These two factors have led the facility designers at the NTS to select the technology of heat pumps as the principal means of providing air conditioning for new facilities.

This chapter discusses the installations of heat pumps at the NTS over the past 5 years. Air-to-air heat pumps installed at the NTS include split systems, single package systems, and through-the-wall systems. A large central plant closed cycle water-to-air heat pump system is also described. The results of a recent analysis of an open cycle water source heat pump system are also presented.

NEVADA TEST SITE

The NTS is located approximately 65 miles northwest of Las Vegas, Nevada, in a high desert climate. The principle elevation of the NTS is between 3,500 and 4,000 feet and is composed of over 1,350 square miles of open desert. Weather data provided by the National Weather Service represent a typical meteorological year.

Temperatures range from winter lows of 5°F to summer highs of 105°F. Daily ranges as great as 40°F are very common throughout the year. Relative humidity levels tend to be low, even during the winters. The NTS ASHRAE design temperatures at 2.5 percent are 98°F DB/64°F WB and 15°F DB. Heating degree days total 5,500 with approximately 160 days a year below 32°F and some 65 days a year above 90°F.

HVAC equipment is selected on the basis of somewhat limited energy sources available to the NTS. Natural gas is not currently available, the nearest pipeline being over 60 air miles away. At the present time, no coal-burning facilities exist at the NTS and therefore no maintenance capabilities are available. Coal could probably be obtained, but emission controls for any required storage areas as well as for the burning equipment would be cost prohibitive. Fuel oil (No. 2) is available to the NTS; however, due to transportation and delivery schedule problems beyond the base camp area, fuel oil is not generally considered for use at NTS locations other than Mercury. Propane (LPG) is also available to the NTS, and like fuel oil is generally considered only for the Mercury area for the same delivery problems. Electrical energy is the prime choice of the nonrenewable supply options from the standpoint of availability and capital cost associated with utilization for HVAC equipment.

AIR SOURCE HEAT PUMPS

Air source heat pumps are self-contained heating and cooling units which rely on the ambient air as the heat source in the heating mode and as the heat sink in the cooling mode. After a false start in the 1950s, air-to-air heat pumps are now a proven and reliable technology. Modern air-to-air heat pumps are available with coefficient of performance (COP) in the range of 2.8 to 3.1 in capacities to 10 tons. COPs are normally listed for equipment operation at 47°F outside air temperature. Air is the most common heat source for heat pumps because of its availability, abundance, and low cost. The major drawback to using air as a heat source is its temperature fluctuation. In air-to-air heat pumps, as the ambient temperature drops, the COP also drops, which has the effect of decreasing the heating capacity. As the heating capacity is being reduced, the heating demand of the facility is increasing. This mismatch widens until the heat pump becomes ineffective and electric resistance backup heating coils must be used.

One current manufacturer provides an air-to-air heat pump that lists a COP above 1.1 at -10°F outside air temperature. Direct electric resistance heating has a COP of 1.0 if no supply fan is utilized. In those cases where forced air heat is necessary, the energy consumed by the fan must be considered, which has the effect of lowering the heating COP below 1.0 for electric heat.

At the NTS, air-to-air heat pumps are utilized for new construction to solve two problems. The first is the cost of energy for heating. Both fuel oil and propane are slightly less expensive than electricity at the NTS on the basis of dollars per million Btu, but the initial installation costs for tankage and piping coupled with maintenance and delivery cost result in the air-to-air heat pump being most life-cycle cost effective. The second is ease of zoning. Many of the buildings constructed at the NTS are pre-engineered metal buildings that tend to be rectangular (40 x 100 is typical) and are best treated as three cooling zones. By installing three split system air-to-air heat pumps, ductwork is minimized and interior comfort is increased.

A large number of dormitory rooms are provided by the NTS in two separate base camp areas. Each room is equipped with a through-the-wall air-to-air heat pump to provide individual control to the dormitory occupants.

WATER SOURCE HEAT PUMPS

Water source heat pumps are self-contained heating and cooling units which rely on water as the heat source in the heating mode and as the heat sink in the cooling mode. Air is used to transmit this heat to or from the conditioned space. A common piping loop is necessary to connect individual units within the system. Water is circulated continuously in this piping loop and heat is either extracted from the water or added to it, depending on whether the load requirement calls for cooling or heating.

Two important engineering advantages exist for water source heat pumps; namely, the COP in the heating mode and the Energy Efficiency Ratio (EER) in the cooling mode remain relatively constant due to the near constant temperature of the source water and all the required piping for water source heat pumps is contained within the building. This piping will require no insulation since the temperature of the circulating water is never cold enough to cause condensation nor warm enough to result in any significant heat losses.

When considering water-to-air heat pumps, two distinct systems are available.

Closed Cycle Water Source Heat Pumps

The closed piping loop in closed cycle heat pump systems is initially filled with a fixed volume of water and this water is used continuously.

Certain disadvantages are apparent with closed cycle heat pump applications.

- In closed cycle heat pump systems, to maintain the proper temperatures in the water loop, the water must be conditioned. Heat input is necessary to maintain temperatures above the minimum and a heat rejector (cooling tower) to eliminate temperatures above the high end of the acceptable range.

- The costs associated with the input heater and cooling tower will substantially raise the initial cost of this heat pump system over that for an air-to-air heat pump system.

- The costs of operating the closed cycle heat pumps could be higher due to the increased energy requirements associated with the water loop circulation equipment.

There are, however, many advantages associated with the use of closed cycle heat pump systems.

- A storage tank can be incorporated into the closed cycle heat pump system. A storage tank installed in the condenser side of the piping loop has the capability of storing heat.

Solar energy can be incorporated into this design as the heat source, reducing the need for the other energy inputs to the water loop.

- In closed cycle systems, the water can be treated to prevent corrosion/erosion that can eventually wear out the piping of the system.

Open Cycle Water Source Heat Pumps

Open cycle water source heat pumps require a water supply piping circuit and a return piping circuit. A constant source of water is necessary for open cycle applications. With open cycle water source heat pump applications, many water source options are available. One option is a well-to-well open cycle. It is necessary that two wells be drilled with this application. One well serves as the supply water for the heat pump piping circuit from which water will be pumped through the system and reinjected into the water table through the second well.

NTS OPEN CYCLE STUDY

Some major problems were discovered as the result of a recently completed study of open cycle water source heat pump applications for the NTS.

All of the water used at the NTS is locally pumped from a depth of over 800 feet. The study looked at three 4-ton heat pumps as the application with the following results.

In this three-heat-pump application, 45 gallons per minute of water are circulated through the system whenever the heat pump is at full load conditions. Equivalent full load hours (EFLH) are the calculated annual hours during which the system will be operating at full load. During the cooling season, there are 758 EFLH and 969 EFLH during the heating season. From these figures, the annual amount of water required for this heat pump application is 4,735,800 gallons.

The drilling of two wells for heat pump applications in new buildings is cost prohibitive. The cost of a water drilling rig is $20,000 per day for maintenance and crew. For a 9 7/8-inch-diameter well with 7-inch casing, the cost of the pump and casing is estimated to be $80,000. The total cost to drill one well 850 feet deep is $280,000. For this particular study, a 25-horsepower pump was required to draw the water from the supply well and a 50-horsepower pump to reinject the water into the return well. The cost of these pumps substantially increased the initial cost of this system. There was also increased operation and maintenance costs.

To avoid the high cost of wells, a second type of open cycle was envisioned. It was assumed that water could be withdrawn from the main community water distribution system, passed through a double-walled heat exchanger, and returned to the community water distribution system at some point remote from the withdrawal point. The distance between these two points has a significant installation and operating cost, and the negative impact of heat removal must be considered.

For example, if 1,000 feet of 6-inch water main were involved in an application, approximately 1,500 gallons of water are in the main. This calculates into 12,500 pounds of water. The flow rate in the heat pump piping loop is 45 GPM or 375 pounds of water per minute of circulating water. It takes 33 minutes to circulate the total volume of water through one cycle. With a COP of 2.6 and the capacity of 49,800 Btu per hour of the heat pump system, it is estimated that 29,800 Btu will be extracted from the water. The water temperature in the community system is relatively constant at 71°F. Since the heat capacity of water is 1 Btu per degree F, the available energy in the piping loop in the 40°F temperature range between 71°F and freezing is approximately 500,000 Btu of stored energy. In 16.5 hours, or 30 cycles, all the heat from the water will be extracted, causing the water to freeze.

The final conclusion is that open cycle water source heat pumps are not practical at the NTS or in any other locale in which the water table is very deep.

CLOSED CYCLE APPLICATION

Three new dormitories are under construction at the NTS at the present time. They are sited in such a fashion that two additional dormitories can be added in the coming years. Each dormitory is designed around 53 sleeping rooms on two stories. Water sourc heat pumps were selected for the full five-dormitory complex. A centrally located utility building will house an oil-fired boiler to provide necessary heat to a large storage tank maintained at 100°F. Should the price of oil increase significantly, it is anticipated that a low-grade solar energy collection system could be retrofitted. The utility building also houses a cooling tower to dump excess heat from the heat pumps. The five-dormitory complex is serviced with a two-pipe circulating water loop maintained between 65°F and 85°F. During the transition seasons of the year, it is anticipated that the entire complex will function as a heat pump; i.e., heat will be pumped from one part of the complex where it is excess to another area where it is needed. No heat is expected to be added by the boiler nor rejected via the cooling tower during the transition seasons.

FUTURE OUTLOOK

Current planning with regard to heat pump strategy at the NTS is to utilize closed cycle water source heat pumps for all new building complexes for which a central utility building can be economically justified.

For new buildings operating only during the normal work schedule (7 a.m. to 5 p.m.), split system air source heat pumps will be considered first. For single room applications, both dormitory and office type, through-the-wall air source heat pumps will normally be considered.

SECTION III
THERMAL STORAGE

6

Thermal Storage Options For HVAC Systems

B.N. Gidwani, M.G. Mirchandani

INTRODUCTION

With the escalating cost of electricity and high demand charges levied by utility companies, thermal storage is rapidly becoming a widely recognized method to lower cooling costs.

In conventional cooling systems, all equipment and piping is sized to meet peak instantaneous load. These systems would run at peak load during design conditions to satisfy existing loads. The concept of thermal storage as applied to cooling systems is that instead of using equipment designed to handle peak loads, a smaller unit is installed. This unit would operate throughout the day storing additional capacity during off-peak hours and using the stored capacity during on-peak periods. This results in smaller equipment, and smaller pipe and duct sizes. In addition, by producing cooling capacity during off-peak hours, the kW is shifted from on-peak periods resulting in demand charge savings. Also, many utility companies have time-of-use or time-of-day rates in which nighttime off-peak energy consumption is cheaper than other periods. Therefore, thermal storage would also save energy charge as expressed in cents/kWh.

Buildings with sharp load profiles are ideal applications for thermal storage. There are two types of storage systems: full storage and partial storage. In a full storage system, the chiller runs during off-peak periods to maximize the savings in on-peak period demand and energy consumption. During the on-peak period, the storage system needs to be sized to handle the entire on-peak cooling load.

This system has a higher initial cost due to larger chiller and storage system as compared to the partial storage system in which the chiller continues to run during on-peak periods supplementing the storage capacity. Before choosing between the two types of storage systems, an economic assessment should be made considering the initial equipment cost and the annual energy costs.

There are three storage mediums commonly in use today: Chilled Water Storage, Ice Storage, and Salt Storage.

This chapter presents the details of each of the above three storage mediums, their relative advantages and disadvantages, applications and cost considerations.

CHILLED WATER STORAGE

This system uses one or more tanks located in the cooling system as shown in Figure 6-1. Chilled water is generated at night by the chiller and flows through the tank, thus charging it. During on-peak period, warm return water enters the tank, displacing chilled water into the system. The principle advantage of the chilled water system is that it is easily retrofitted to existing chillers and does not sacrifice chiller efficiency. However, this medium of thermal storage requires a very large space since it utilizes sensible heat for storage. For a system operating at a 15°F temperature difference, the storage would be 15 Btu per pound of water. The major problem with existing chilled water systems has been in the area of blending the chilled water with the warm return water. Several techniques ranging from membranes to multi-tanks have been used to prevent temperature blending in the storage tank. Membranes are a source of maintenance problems whereas multi-tank systems result in high initial costs.

ICE STORAGE SYSTEM

Ice storage systems fall into two different categories. First there are the static systems in which ice is formed on the evaporator coils inside an ice bank. The storage capacity of the ice bank is a function of the ice thickness. The ice bank generally consists of a serpentine pipe coil submerged in water in an insulated box. There are no moving parts within the ice bank. Design evaporating temperatures range from 10°F to 25°F depending on desired ice thickness, ice build-time, and heat transfer area.

The second kind of ice-storage system is the dynamic system in which ice is produced in crushed or chunk form and fed into large storage tanks. Water is circulated through the tank during on-peak periods to produce chilled water for cooling. Static systems are more popular since they are available as factory-assembled packaged units which provide ease of installation and a lower initial capital cost.

Since the latent heat of fusion is used for ice-storage, a pound of ice stores 144 Btu's of cooling energy as compared to 15 Btu's per pound for a chilled water system. This results in a storage capacity of about ¼ that required for chilled-water. Other advantages include less weight, easier controls than water-storage systems and greater reliability. A disadvantage of ice-storage systems is that they cannot be easily retrofitted to existing chillers because existing chillers are typically designed for about 42°F water, whereas, evaporator temperatures required for ice-storage systems are 10 to 25°F. Another aspect of ice-storage systems that warrants attention is that the lower evaporator temperature results in a slightly higher kW/ton for the system causing some penalty in cooling efficiency. Figure 6-2 illustrates a typical ice-storage system.

SALT STORAGE

This system utilizes salt hydrates which are mixtures of inorganic salts and water that are capable of freezing and melting at a preselected temperature. Salts are available that are capable of the above phase change at 47°F. Typically, the salts are sealed in plastic containers which are stacked in layers in a large tank. There are gaps between the containers which allow water to flow between the layers. During off-peak periods, chilled water is passed through the tank freezing the salts in the containers. The circulating water does not come in direct contact with the salt at any time. During the "melt-down" cycle (on-peak periods), the warm water from the building circulates through the tank, melting the salt and in turn is cooled to meet the cooling load. The latent heat of fusion of the salt-solution is about 40 Btu/lb. The tank size required is about 1/3 that required for water-storage. Other advantages of salt-storage systems are that they do not sacrifice chiller efficiency and they are easily retrofitted to existing chillers. Since salt-storage is a relatively newer technology, it does not have a proven storage track record unlike the other two storage mediums. Moreover, since the lowest water temperature possible is 47°F, humidity requirements of the building should be considered before specifying salt-storage. Also, the added cost of the salt containers should be considered in the economic analysis. Figure 6-3 illustrates a typical salt-storage system.

SIZING THERMAL STORAGE SYSTEM

The first step in sizing a thermal storage system is to develop the peak-day load profile. This can be done utilizing chiller logs where available, or using load estimation software, or computer simulation of load based on ambient conditions. Table 6-1 presents a typical peak day load profile calculated using an hourly computer simulation model. No chiller logs were available, but it is known that

- Maximum Load = 674 Tons
- Design Day Temperature = 93°F DB
- Minimum Load = 236 Tons
- Temperature at Minimum Load = 40°F DB

The cooling load in tons is

Minimum Load + (Maximum Load − Minimum Load) x

$$\frac{\text{(Hourly Temp.} - \text{Temp. at Minimum Load)}}{\text{(Design Day Temp.} - \text{Temp. at Minimum Load)}}$$

Having obtained the peak-day load profile, the next step is to size the thermal storage capacity. The total daily cooling capacity required is 14,102 ton-hours. During summer months (May 1 to September 30) the on-peak hours on a weekday are effective from 12:30 p.m. to 6:30 p.m. If both chillers were to operate only during partial-peak and off-peak hours–6:30 p.m. to 12:30 p.m. (18 hours)–the available capacity is calculated as follows:

18 Hours x 2 x 337 Tons = 12,132 Ton-Hours

However, there is still a remaining cooling capacity of

14,102 − 12,132 = 1,970 Ton-Hours

which will have to be generated during on-peak hours on a peak summer day. This load can be met by running one chiller for

$$\frac{1,970 \text{ Ton-Hours}}{337 \text{ Tons}} = 5.85 \text{ Hours}$$

Say 6 hours from 12:30 to 6:30 p.m.

Therefore, on a peak summer day the chiller operation can be summarized as follows:

- 6:30 p.m. to 12:30 p.m. (18 hours) − both 337 ton chillers run at full load.
- 12:30 p.m. to 6:30 p.m. (6 hours) − one 337 ton chiller runs at full load.

Even on a peak summer day, thermal storage helps in shifting 368 kW (one chiller) from on-peak to off-peak and partial-peak hours in the example.

Having computed the peak summer-day load profile and the proposed chiller operation, the size of the thermal storage system can be estimated. From 6:30 p.m. both chillers operate. Every hour the chiller runs (6:30 p.m. to 12:30 p.m. in the above example), it produces 674 tons of cooling and the air conditioning load required for that hour is used. At 1900 hours (6:30 p.m. to 7:30 p.m.), 674 tons of cooling is produced and 632.7 tons of air conditioning are required. The available capacity at the end of 1900 hours is

674 − 632.7 = 41.3 Tons

At the beginning of 2000 hours (i.e., at 7:30 a.m.), 41.3 tons of cooling capacity is available from the previous hour; 674 tons are produced till 8:30 p.m. and 616.15 tons are used. Therefore, the available capacity at the end of 2000 hours (i.e., at 8:30 p.m.) is

41.32 + 674 − 616.2 = 99.2 Tons

Table 6-2 presents an hourly dynamic profile of the available capacity at the end of each hour. From 12:30 p.m. to 6:30 p.m. only one chiller is operating. As can be seen from the table, the thermal storage system should be capable of storing at least 1992 ton-hours which is the maximum available capacity that would occur on a peak summer day.

SAVINGS CALCULATIONS

Figure 6-4 shows the peak day load profile and the proposed chiller operation with thermal storage. The energy savings will result from the following:

- Demand Savings =

    Σ (kW Shifted)$_i$ x (\$/kW)$_i$

    and,

- Energy Charge Savings =

    Σ (Existing kWh Charge)$_i$ −
    (Proposed kWh Charge)$_i$

    Where,

    i = Each Month

ECONOMIC ANALYSIS

To assure that thermal storage is an attractive project for a particular installation, a complete life cycle cost analysis should be performed. In addition to the energy savings there are a number of other factors that need to be addressed. The various factors to be considered in the life cycle analysis are:

- Capital cost of the thermal storage system.
- Any rebate or incentives that can be obtained from the Utility.

Thermal Storage Options for HVAC Systems

- Avoided Cost, if any, for a new or backup chiller.
- State energy tax credit savings, where applicable.
- Energy and demand savings.
- Annual operating and maintenance costs.

Figure 6-5 illustrates the various factors involved in the life cycle cost analysis.

In summary, it is imperative for the designer of any thermal storage system to consider the various storage technologies and systems, and to evaluate them for the particular application as discussed in this chapter.

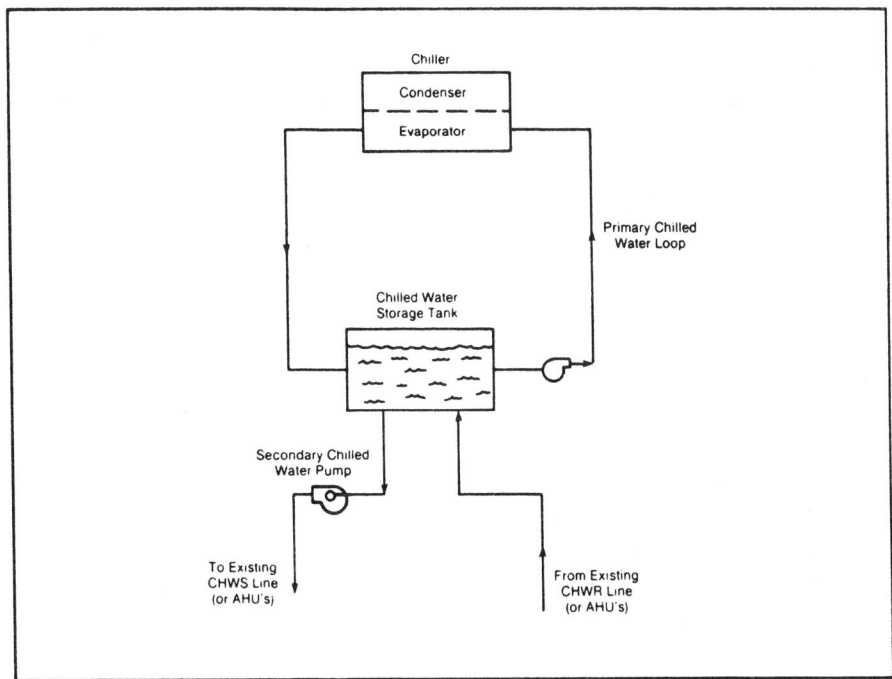

Figure 6-1. Schematic of chilled-water storage system.

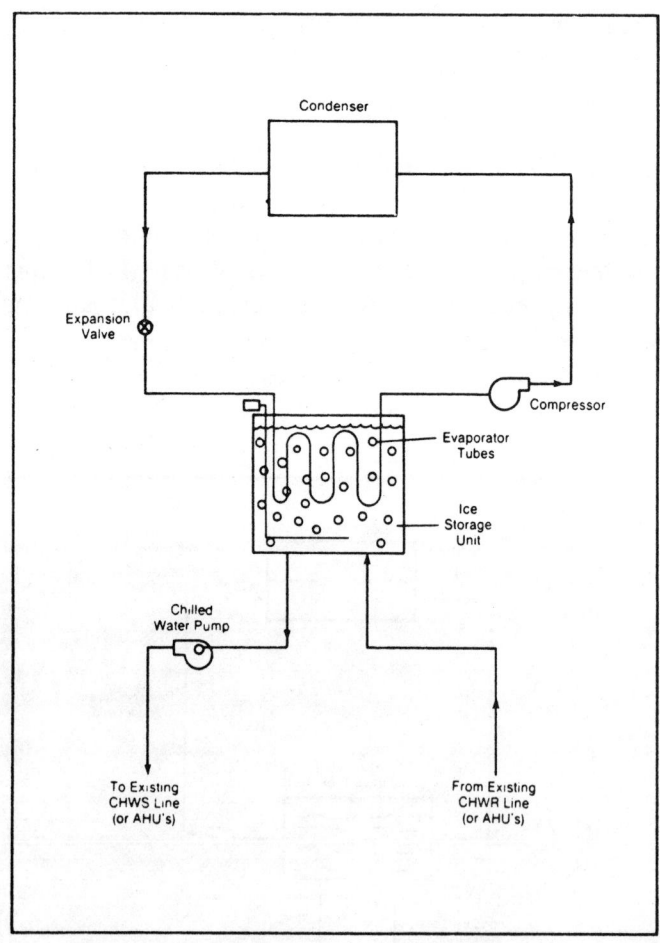

Figure 6-2. Schematic of DX ice-storage system.

Thermal Storage Options for HVAC Systems

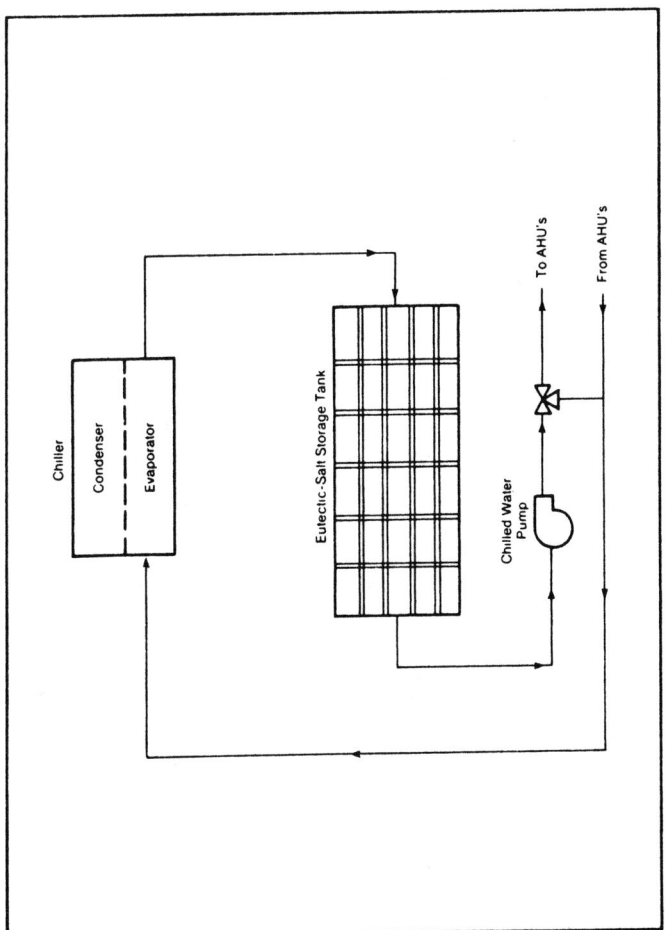

Figure 6-3. Schematic of salt-storage system.

HOURS	TEMPERATURE DBT	TEMPERATURE WBT	COOLING CAPACITY (TONS)
000	77.00	66.00	541.77
100	75.00	66.00	525.25
200	74.00	65.00	516.98
300	73.00	65.00	508.72
400	73.00	65.00	508.72
500	74.00	65.00	516.98
600	75.00	65.00	525.25
700	77.00	66.00	541.77
800	79.00	66.00	558.30
900	81.00	67.00	574.83
1000	83.00	67.00	591.36
1100	85.00	68.00	607.89
1200	88.00	69.00	632.68
1300	90.00	70.00	649.21
1400	92.00	70.00	665.74
1500	93.00	70.00	674.00
1600	92.00	70.00	665.74
1700	91.00	69.00	657.47
1800	90.00	69.00	649.21
1900	88.00	69.00	632.68
2000	86.00	68.00	616.15
2100	84.00	68.00	599.62
2200	82.00	67.00	583.09
2300	79.00	67.00	558.30

Total Peak Day Ton-Hours: 14,101.70

- Temperature at Minimum Load 40°F DB
- Temperature at Design Day   93°F DB
- Minimum Load   236 Tons
- Maximum Load   674 Tons

Table 6-1. Typical calculation of peak day load profile by an hourly simulation model.

Thermal Storage Options for HVAC Systems

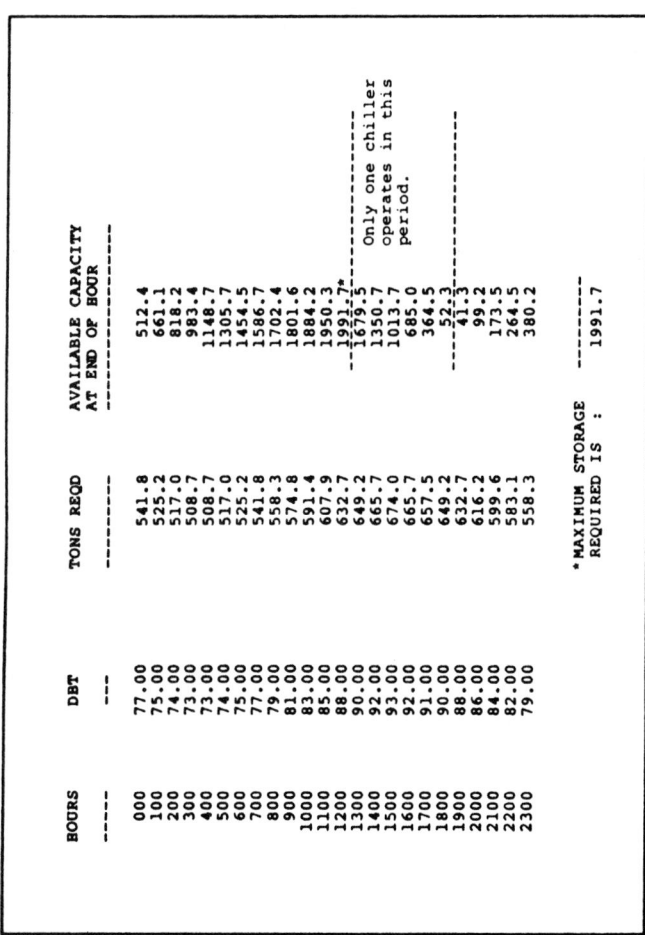

HOURS	DBT	TONS REQD	AVAILABLE CAPACITY AT END OF HOUR	
000	77.00	541.8	512.4	
100	75.00	525.2	661.1	
200	74.00	517.0	818.2	
300	73.00	508.7	983.4	
400	73.00	508.7	1148.7	
500	74.00	517.0	1305.7	
600	75.00	525.2	1454.5	
700	77.00	541.8	1586.7	
800	79.00	558.3	1702.4	
900	81.00	574.8	1801.6	
1000	83.00	591.4	1884.2	
1100	85.00	607.9	1950.3	
1200	88.00	632.7	1991.7*	
1300	90.00	649.2	1679.5	Only one chiller operates in this period.
1400	92.00	665.7	1350.7	
1500	93.00	674.0	1013.7	
1600	92.00	665.7	685.0	
1700	91.00	657.5	364.5	
1800	90.00	649.2	52.3	
1900	88.00	632.7	41.3	
2000	86.00	616.2	99.2	
2100	84.00	599.6	173.5	
2200	82.00	583.1	264.5	
2300	79.00	558.3	380.2	

*MAXIMUM STORAGE REQUIRED IS : 1991.7

Table 6-2. Sizing of thermal storage capacity.

118   OPTIMIZING HVAC SYSTEMS

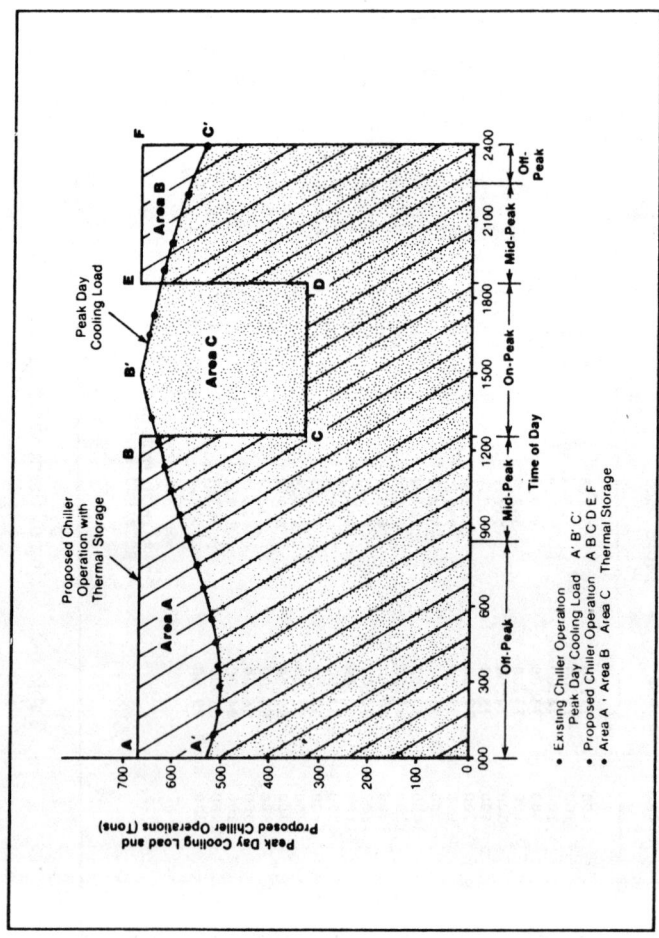

Figure 6-4. Peak day cooling load profile and proposed chiller operation.

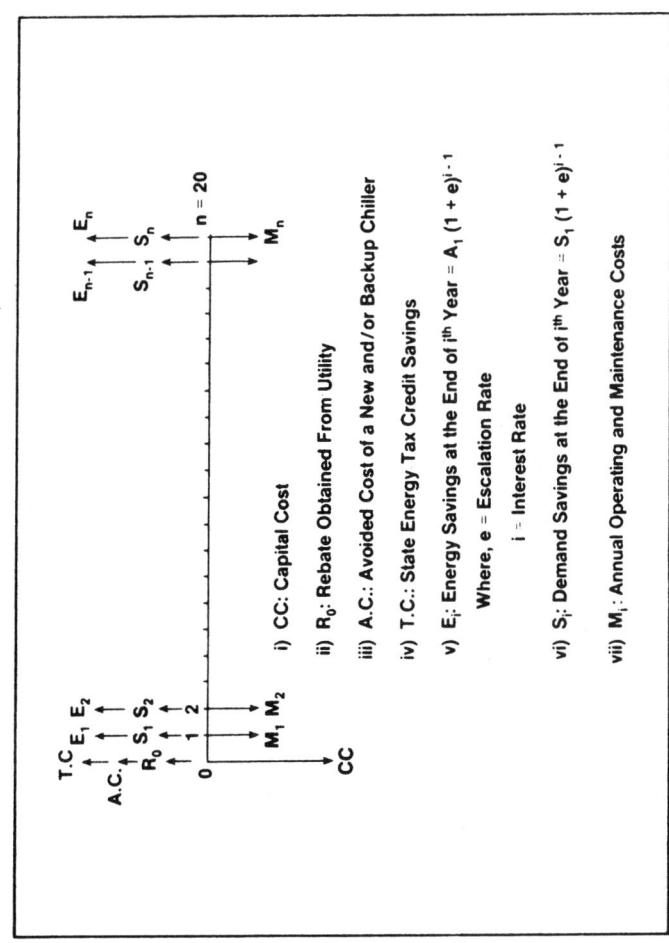

Figure 6-5. Life cycle costing for various thermal storage options.

7
Analysis of Thermal Storage Options

B. Bechard, E.A. Evans

INTRODUCTION

A thermal storage system can be compared to a rechargeable flashlight battery. The thermal battery stores heating or cooling energy for use at a later time. The storage medium may be water, ice, eutectic salts, or other media such as rocks. The storage medium is heated or cooled (charged) during periods of low energy usage and is discharged during periods of high energy usage. The cycle is then repeated. Most thermal storage systems operate on a 24-hour cycle; however, weekly or yearly cycles are not uncommon. Thermal storage systems provide a means to decouple the use of thermal energy from the thermal load being served. Thermal storage options used for chilled water storage will be analyzed in this chapter.

THERMAL STORAGE BENEFITS

Thermal storage systems provide many benefits, including reduction of utility costs and increased reliability. A summary of these benefits is provided below.

- Lower Demand Costs: The peak electrical demand for most facilities coincides with the peak cooling demand. The peak electrical demand and the demand charge for the peak months can be reduced considerably by storing cooling energy during off-peak hours. Depending on the utility rate structure, this might also reduce the demand charges for each month of the year.

- Lower Electric Consumption Costs: Many utility companies offer reduced electric rates during off-peak periods. A thermal storage system allows the owner to take advantage of these rates.

- Reduced Equipment Sizes: A building initially designed with a thermal storage system may have a smaller chiller installed. Because the off-peak periods are generally longer than the on-peak periods, a smaller chiller can run longer to generate the same cooling energy. If the storage medium is ice, the temperature differences in the system may be increased, and the sizes of the air and water distribution system can therefore be reduced. This results in lower initial costs for the equipment and lower operating costs.

- Increased Reliability: Facilities with large computer centers which must be air conditioned will benefit from thermal storage. Critical computers usually have a UPS or emergency generator to provide power during short power outages. Often, the chiller serving the computer area does not have emergency power. A thermal storage system can provide emergency cooling for computer centers and thereby increase their reliability and operating time.

- Increased Efficiencies of Chiller Systems: A chiller sized to meet peak building loads will operate below its peak efficiency most of the time. A chiller which is charging a thermal storage system can operate at peak efficiency all of the time. When the storage tank is fully charged, the chiller can be shut down.

CANDIDATES FOR THERMAL STORAGE

Thermal storage systems are not economical in every situation. A careful engineering analysis of each specific case is required to determine its economic benefits. The following criteria should be considered before detailed analysis is completed. If the project meets these criteria, the potential for an economical project is high.

Load Profile

The load profile for the facility being considered for thermal storage must have peaks and valleys. If the load is constant for a 24-hour period, there is little benefit in providing thermal storage. A typical office building has a load profile which is ideal for thermal storage. Figure 7-1 shows the breakdown for the various load elements for a typical office located in Philadelphia. The chiller load, lighting load, and office equipment load vary considerably during the 24-hour period. The chiller load also varies seasonally and adds significantly to the daily electric peak.

Electric Utility Schedules

Some electric utility schedules are better suited for thermal storage than others. More and more utility companies are penalizing the electric users with high demand charges during peak times and rewarding the off-demand users by lowering kWh costs and demand charges during the off-peak hours. This type of utility contract provides an excellent opportunity to maximize thermal storage benefits. A high-demand charge during peak use hours usually is enough incentive to economically justify a thermal storage system. A lower kWh cost during off-peak hours is icing on the cake. Some utility companies are offering large incentives to facilities who install peak avoidance systems. This could amount to $200/kW-$300/kW in a lump sum amount for avoided demand. These incentives can significantly impact the overall economics of a thermal storage system.

Table 7-1 provides a summary of typical utility schedules for various areas of the country. Note that Philadelphia has a high demand charge but very little difference between on- and off-peak kWh costs. San Diego has a lower demand charge but a greater difference between on- and off-peak rates.

ANALYSIS OF THERMAL STORAGE OPTIONS

Options to consider when designing a thermal storage system include storage system type and size, and chiller system type and size. The advantages of the various options are presented below.

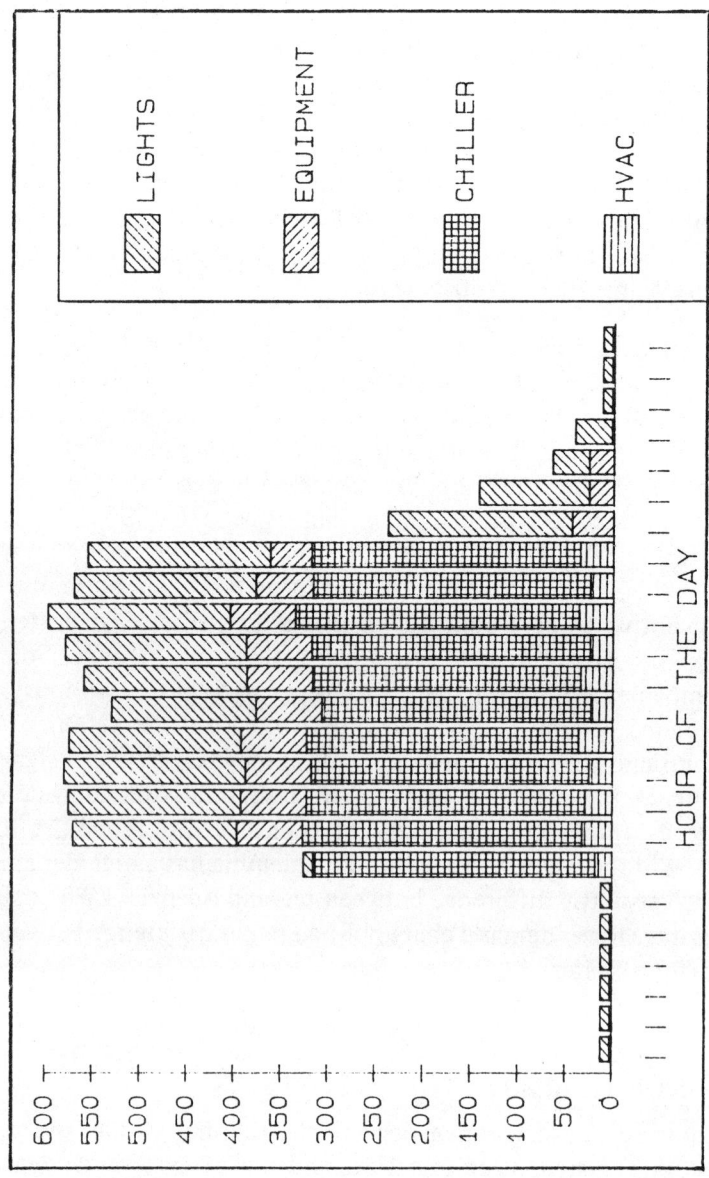

Figure 7-1. Hourly kW demand profile for peak cooling day.

Analysis of Thermal Storage Options

City	Effective Demand Charge $/kW	On Peak $/kWh	Off Peak $/kWh	Demand Ratchet	Peak Avoidance Incentives
Denver	6.15	0.0367	0.0248	None	None
Phil.	13.45	0.0433	0.0355	80% of summer peak	$0.0021/kWh off-peak credit
San Diego	7.81	0.12934	0.05898	None	$225/kW up-front cash

Table 7-1. Electric utility contracts for various locations.

Storage Media

Three types of media are commonly used in thermal storage systems: water, eutectic salts, and ice. All three have advantages and disadvantages. Some are more suited to new construction, and some are more suited to retrofit applications.

Water systems are the most common and, in a retrofit application, usually provide the best economics. Water systems operate at temperatures which allow use of existing chiller and air-side equipment without any changes. A large volume storage tank is required; this could be a problem where space is limited.

Several schemes have been developed for water storage which separate the cold supply water from the warm return water. The simplest but costliest method is to provide two tanks; one for cold water storage and the other for warm water storage. During the discharge mode, the water is pumped from the cold tank to the building load and back to the warm storage tank.

Another method is to provide a series of six to eight smaller tanks. (See Figure 7-2.) One tank is initially empty. Chilled water is pumped from the first tank to the building load and then returned to the empty tank. When the first tank is empty, it is then used as the warm water storage tank. This method, however, requires a costly valving system and a complicated control system to open and close valves. It does reduce the space required for storage when compared with the two-tank systems.

126   OPTIMIZING HVAC SYSTEMS

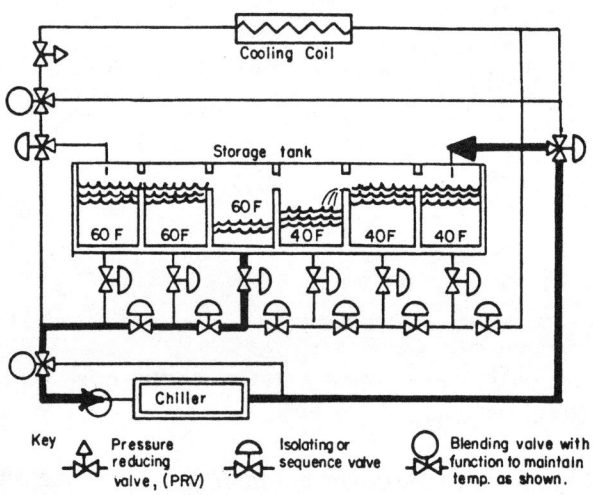

a) Charging mode

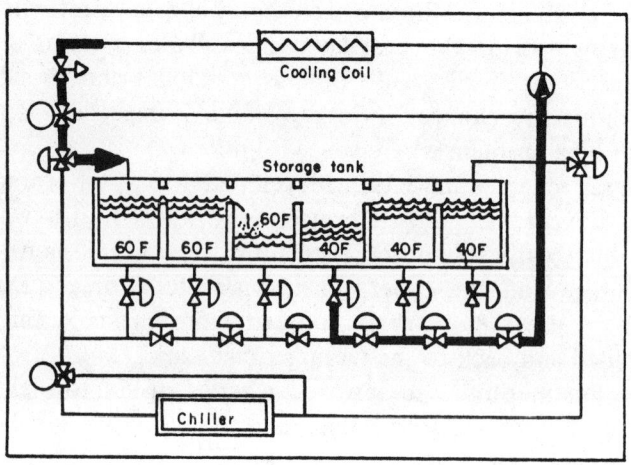

b) Discharge mode

Figure 7-2. Chilled water storage: multiple tank system.

Analysis of Thermal Storage Options 127

Figure 7-3 shows a third method of chilled water storage. This method utilizes a single storage tank with a flexible bladder separating the warm and cold water. As water is pumped from one side of the bladder and returned to the other, the bladder moves up and down. This method provides the most economical approach to chilled water storage.

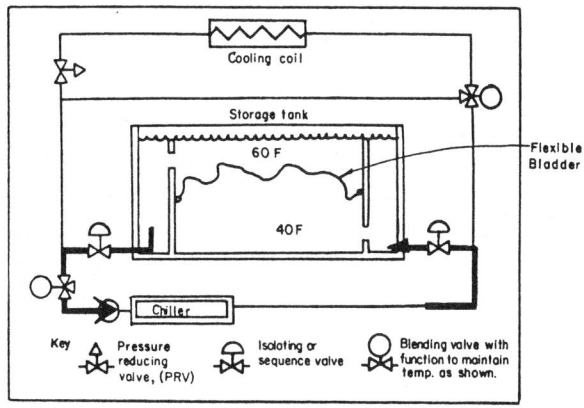

a) Tank charging mode

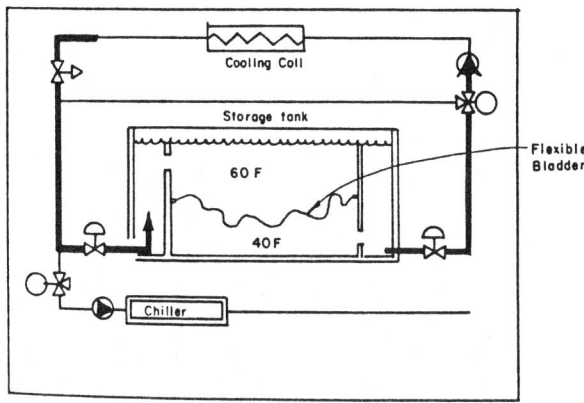

b) Tank discharge mode

Figure 7-3. Chilled water storage: single tank with flexible bladder.

Using ice as a storage medium can reduce the size of the storage tank considerably. One pound of water will store 16 Btu of cooling energy when cooled from 60°F to 44°F. If one pound of water is cooled from 60°F to 32°F and changed to ice, it will store 172 Btu of cooling energy. In theory, ice would utilize about 10% of the storage volume needed by a water system to store the same amount of cooling energy. In practice, however, only about 50% of the water in a tank can be frozen.

Unfortunately, the cost of the ice storage system will not be reduced in proportion to its size. The cost of ice builders, additional piping, and brine solutions offset the cost of the smaller storage tank. Direct expansion systems reduce the cost somewhat, but these systems are very difficult to operate and maintain, and are not recommended. Figure 7-4 shows an ice storage system which utilizes a brine solution and a heat exchanger to separate the chiller water loop from the brine.

While the ice storage method has the advantage of a smaller storage volume, it has the disadvantage of requiring considerably lower compressor operating temperatures. This decreases the compressor operating efficiency considerably. These losses can be offset by other factors which will be discussed further on.

The use of eutectic salts as storage medium provides the advantage of low storage volume, similar to ice storage systems, but without the loss of compressor efficiency.

Eutectic salts are similar to the "blue ice" in packages which are available in sporting goods stores and commonly used to keep picnic lunches cold. These salts have the advantage of freezing at temperatures between 40°F and 50°F. The phase change provides about 40 Btu/lb of energy storage. This is not as good as ice but much better than plain water. Eutectic salts come in plastic containers which are stacked inside a storage tank. The chilled water is circulated directly over the containers, so no heat exchangers are required. The only drawback with this system is the initial cost of the eutectic salt. Table 7-2 provides a summary of the cost for the three storage systems discussed, each sized for 4400 ton-hours of capacity.

Analysis of Thermal Storage Options

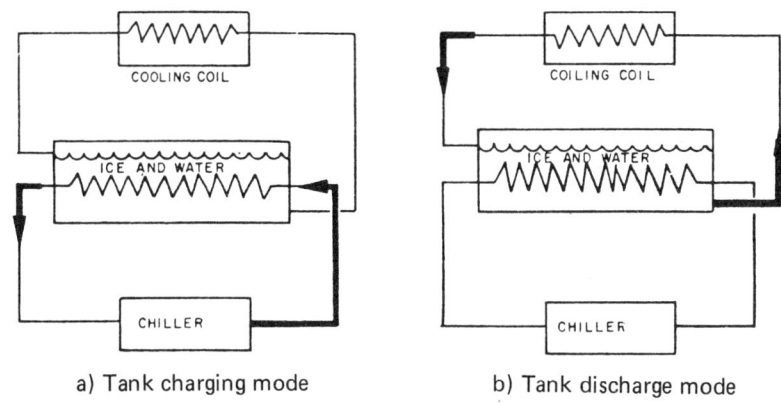

a) Tank charging mode   b) Tank discharge mode

Figure 7-4. Ice storage system with brine and heat exchanger.

ICE STORAGE SYSTEM
    4400-ton-hour insulated storage tanks $355,740
    Pumps   12,376
    Piping   13,817
    Valves, wiring, controls   12,688
    TOTAL $394,621

EUTECTIC SALT STORAGE SYSTEM
    4400-ton-hour insulated storage modules $396,000
    Pumps   13,728
    Piping   18,135
    Valves, wiring, controls   14,300
    TOTAL $442,163

CHILLED WATER STORAGE SYSTEM
    475,860-gallon insulated storage tank $117,168
    Concrete foundation   30,912
    Bladder   41,600
    Pumps   13,728
    Piping   18,135
    Valves, wiring, controls   14,300
    TOTAL $235,843

Table 7-2. Summary of Storage Costs

Chiller Systems

The type of storage medium selected, as indicated earlier, depends on whether the thermal storage system is a retrofit or a new installation. With a retrofit, the chiller and air-side equipment are usually designed for chilled water of approximately 40°F-45°F. The thermal storage system must accommodate these design restrictions. A selection of chilled water storage or a eutectic salt solution would be recommended for a retrofit application.

Using an ice storage system for a retrofit is usually not advisable. A severe penalty would be paid in chiller efficiency by going to the lower temperature. Figure 7-5 shows the reduction of efficiency and capacity which results from lowering the suction temperature of a typical centrifugal compressor.

The fact that the chiller will operate only at night will not appreciably increase the operating efficiency. A common misconception is that the lower night temperatures will reduce condenser water temperatures. The condensing water temperature, however, follows the outside wet-bulb temperature. Figure 7-6 is a graph plotting the outside dry-bulb and wet-bulb temperatures for a typical summer day. The wet-bulb temperature varies little when compared with the dry-bulb swing.

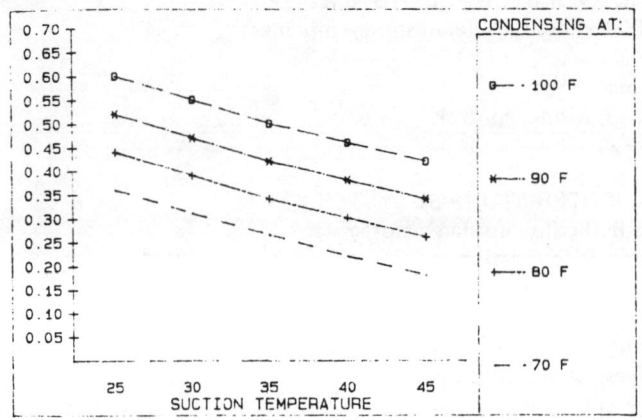

Figure 7-5. Chiller kW/ton versus suction temperature.

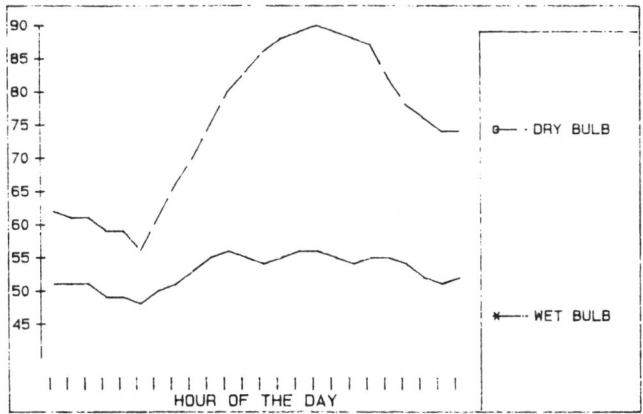

Figure 7-6. Typical hourly profile of summer dry-bulb and wet-bulb temperatures.

The penalty of lower chiller efficiency which results from ice storage systems can be overcome if the ice storage system is designed as part of the original, total HVAC system. The lower chilled water temperatures (32°F-34°F) can provide some additional benefits in the design of the HVAC equipment. The chilled water flows and pumps sizes can be reduced. The air-side equipment can be selected for lower supply air temperatures, reducing supply CFM, duct sizes and fan sizes. This leads to considerable savings in capital cost for HVAC equipment and results in reduced operating costs.

COMPUTER MODELLING

Varying climatic conditions affect the performance of cooling towers. The energy consumption of HVAC equipment is influenced by hour-by-hour changes in solar intensity, ambient wet- and dry-bulb temperatures, and effects of the internal loads of a building due to occupants, lighting, and equipment. Predicting the cost of building operation is further complicated when the utility serving the building is billing for electricity use based on a time-of-day rate structure. Such a combination of variables presents a complex problem to the thermal storage analyst. Hourly computer simulation can make the job easier.

Sophisticated building energy models such as DOE 2.1 have the capability of simulating a wide variety of heating and cooling systems, including thermal storage. Analysis results can be very reliable, but they are only as good as the data input to the program. If you have a good grasp on how the building is currently using energy and can generate a good facsimile with the computer, then your thermal storage results should be reliable. The more you know about current energy use (e.g., annual and monthly consumption, hourly profiles of peak day use), the more accurate will be your base case computer simulation; consequently, your thermal storage system simulation will yield accurate results.

The point of installing a thermal storage system is to save money. Most of the savings are usually realized from avoided demand charges. The rest are due to off-peak electric energy rate differentials. However, as we shall see, this is not always the case. The magnitude of the electric loads that can be shifted to off-peak hours will define peak demand savings for thermal storage systems. The effects of time-of-day energy use can be determined by comparing daily demand profiles of the base case to those of the storage case. The computer can be used to generate these profiles and to calculate electricity use by time of day for every day of the year.

TWO THERMAL STORAGE STRATEGIES

Two strategies for application of thermal storage systems are presented here. The first employs a tank large enough to store cooling energy to handle the entire building load for the peak cooling day. The second employs a tank that will store only enough energy to meet part of the load.

For illustrative purposes, consider this thermal storage candidate: a three-story, 117,000 ft^2, commercial office building located in Philadelphia. After considering first costs and compatibility of alternate storage media with the existing chiller, chilled water was selected as the storage medium. This building is currently served by a variable-air-volume system which is run in the cooling mode 5 months a year, May through September. The chilled water coils are fed by a 400-ton centrifugal chiller operating from 6:00 a.m. to 5:00

Analysis of Thermal Storage Options

p.m. and providing cooling to spaces dominated by general office-type use. The building is occupied from 7:00 a.m. until 6:00 p.m., Monday through Friday. Because the chiller and fans are shut down at night, this schedule is well suited to using the unoccupied hours to generate and store chilled water for the cooling needs of the next day.

The daytime cooling loads are created by various factors: internal heat generated by lighting, office equipment, and building occupants; gains from heat conduction through the building envelope; solar gains through the windows; and outside air introduced into the building by mechanical systems.

Occupant density is approximately 85 ft^2/person. The lighting level is 1.65 W/ft^2, and office equipment uses about 0.75 W/ft^2. Because of the thermal capacitance of the building structure and its contents, heat is stored in walls and furniture during the day. When the mechanical system fans are shut down around 5:00 p.m., ventilation is eliminated. Heat stored in the building structure and contents is released into the building during the night, and temperatures inside the building remain relatively warm despite cooling outside temperatures. Early morning solar gains contribute to the interior heat level. Consequently, cooling is sometimes needed at 6:00 or 7:00 a.m. to bring the interior temperatures of the building down to the comfort range by the time people start arriving for work. Even if cooling loads do not occur before the workday starts, loads are rapidly generated when people arrive, and lights and equipment are energized.

Figure 7-1 represents the peak-day hourly electric load profile for this building before installation of the storage system. The components of the electric load are chiller load, lights, office equipment and HVAC equipment. The chiller load represents about 50% of the peak demand and also represents the load which can be shifted to off-peak, night-time hours by using thermal storage.

Full Storage

The non-chiller electric loads of this building are roughly equivalent to those imposed by the chiller. Without changing the operation of the building, peak demand can be nearly halved by relegating chiller operation to strictly nighttime thermal storage charging. To accomplish this, storage must be sized to accommodate the entire cooling load experienced on the peak day.

The cooling load for the design day defines the ton-hours of cooling that must be delivered by the HVAC system; storage system losses (e.g., tank heat gains, stratification, unrecoverable stored energy) must be accounted for. The achievable temperature drop across the chiller under design day conditions greatly influences the size of the required storage volume.

The design day cooling load for this building (estimated from the data in Figure 7-1) is 3517 ton-hours. Assuming design day storage system losses of 30% and a temperature difference of 16°F, the required water storage volume is 411,121 gallons. The closest standard-size tank is 475,860 gallons.

The existing 400-ton chiller generates chilled water during the off-peak electric period (8:00 p.m. to 8:00 a.m.). This system produces 560 gpm of water cooled an average of 16°F in one pass through the chiller. The water is stored in a single tank with a bladder similar to the system shown in Figure 7-3.

Figure 7-7 depicts the hourly electric peak demand profile for the storage case. The entire chiller load has been shifted to off-peak hours.

CALCULATION OF SAVINGS

As Figure 7-8 demonstrates, the storage system shaves the kilowatts drawn by the chiller and cooling tower from the base case daytime peak and shifts the demands of this equipment to the off-peak period. The DOE 2.1 analysis program can predict the peak demand for each month resulting from this shift as well as the number of kilowatt-hours of electrical energy consumed during the peak and off-peak periods over the course of the cooling season.

The nature of the cost-impact analysis of storage systems requires hour-by-hour determination of electricity use. For this reason, computer models that perform hourly system simulation provide a much more accurate assessment of thermal and economic performance than do other methods. Computerized design tools intended for use in sizing equipment can provide reliable estimates of peak kW

Analysis of Thermal Storage Options 135

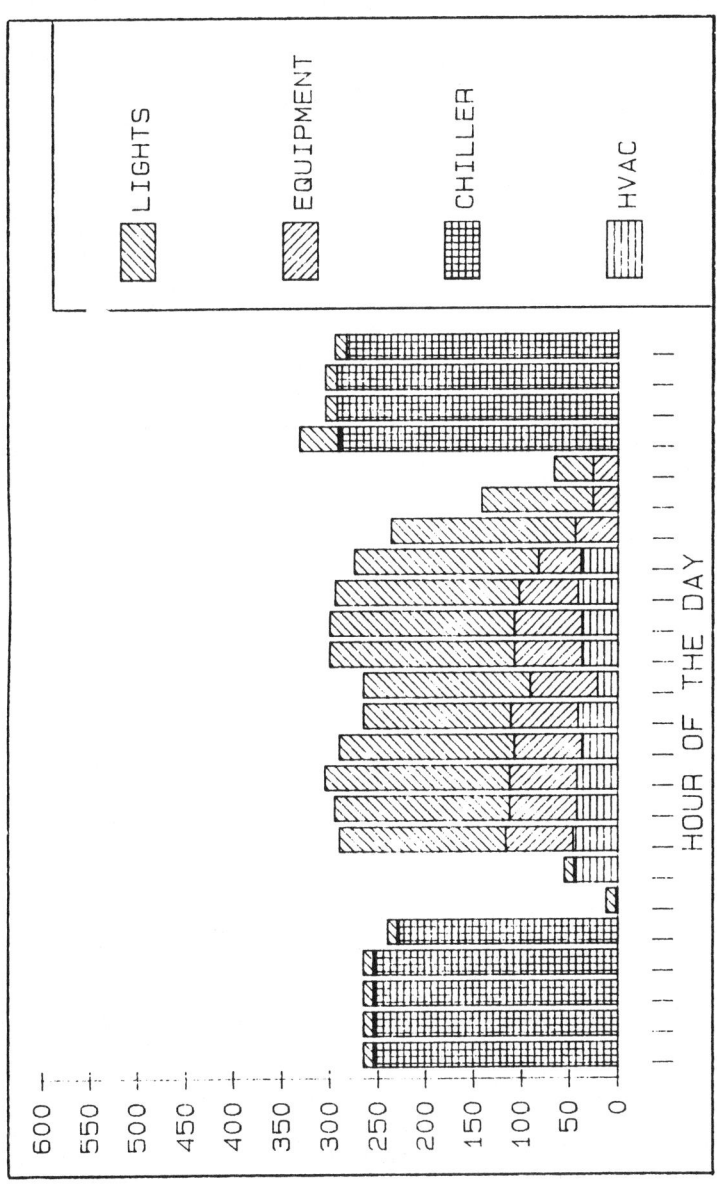

Figure 7-7. Hourly kW demand profile on peak cooling day with full storage system.

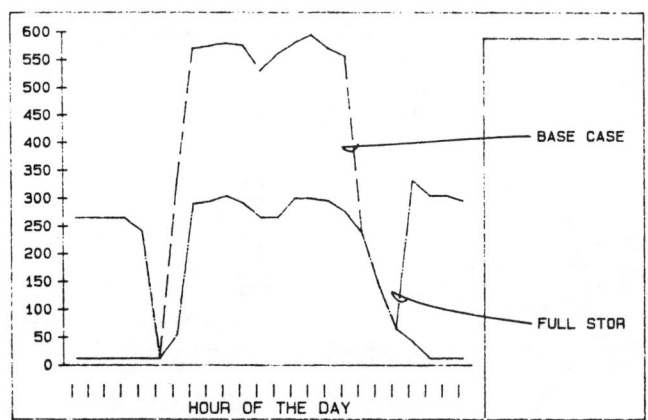

Figure 7-8. Peak kW demand profiles: base case versus full storage.

demand effects of storage systems, but they do not present an accurate picture of on-peak versus off-peak electricity consumption. In certain utility service areas, savings resulting from the shift of energy use to off-peak hours exceed peak kW savings. Therefore, hourly energy simulation is highly recommended.

With monthly peak kW demand data and totals of on-peak and off-peak electricity use in hand for both the base case and the storage case, the energy and dollar savings delivered by thermal storage can easily be determined. Even complex utility rate structures are easily analyzed with the use of appropriate computer software.

Table 7-3 presents the monthly cost savings generated by this storage system. Full storage enables this Philadelphia Power and Light customer to avoid $33,934 in demand charges each year and $3,960 in energy charges. The system will pay for itself in less than 6 years.

Partial Storage

The next case involves a storage system sized to address only part of the building cooling load. The chiller generates chilled water during off-peak hours but only in an amount that covers about half

Month	Full Storage	Partial Storage
Jan	$ 2,993	$ 1,227
Feb	2,993	1,227
Mar	2,993	1,227
Apr	2,993	1,227
May	3,392	1,859
Jun	4,055	1,945
Jul	4,037	2,033
Aug	4,201	1,924
Sep	4,251	2,206
Oct	0	0
Nov	2,993	1,227
Dec	2,993	1,227
TOTAL	$37,894	$17,329

Table 7-3. Monthly cost savings: Philadelphia.

of the design day load. On high cooling-load days, the chiller operates in parallel with the storage tank. During medium cooling-load days, the chiller operates at part-load to augment storage in meeting cooling needs. On low cooling-load days, the storage tank covers the cooling requirements by itself.

The advantage of the partial storage approach is the lower first cost when compared with the full storage system. The tank is roughly half the size of the full storage tank, and the total system cost is about 30% less than for full storage in a retrofit application. Front-end savings are even greater in a new construction project, because the chiller can be down-sized to half the capacity required by full storage or a conventional cooling plant. In this case, system costs are about 40% lower.

To size the partial storage tank and chiller, the design day cooling load is divided by 24 to get a levelized load over a 24-hour period. The resulting tonnage represents the chiller capacity. Ton-hours of storage are then defined by the number of off-peak hours the chiller will operate each night.

For our example building, partial storage volume is 225,045 gallons after accounting for system losses. The closest standard-size tank is 225,582 gallons.

Figure 7-9 depicts the effect of partial storage on the peak electric load profile. As with the analysis of full-storage systems, the computer is an invaluable tool in assessing the merits of partial storage. Again, it is very important to be able to quantify the electrical energy use during peak and off-peak hours.

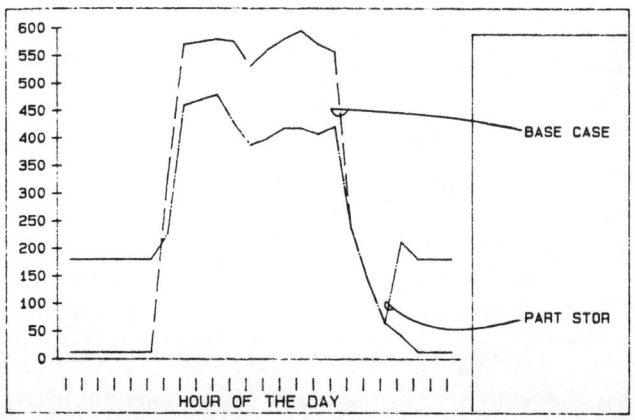

Figure 7-9. Peak kW demand profiles: base case versus partial storage.

Table 7-3 presents the monthly cost savings provided by this partial storage system. Demand charges and energy charges in the amount of $14,903 and $2426 are avoided by using this system. These smaller savings relative to full storage are offset by the 29% lower retrofit system cost to yield a 9-year simple payback.

The ease of analyzing these systems with a fully capable computer model and utility rate calculation package makes optimization of storage systems quick and efficient.

EFFECT OF A DIFFERENT RATE STRUCTURE

To illustrate the impact of the utility rate structure on system economics, full- and partial-storage systems for this same building were modelled in San Diego. San Diego Gas and Electric has rates

Analysis of Thermal Storage Options

that are very favorable for load management systems like thermal storage. In addition, an up-front cash incentive based on the amount of displaced demand makes San Diego a very attractive thermal storage market.

The procedures described previously were employed to size full- and partial-storage systems to address the design day cooling load of 3,069 ton-hours in San Diego.

Figure 7-10 demonstrates the kW demand shaved from the base case operation by a full-storage system. Figure 7-11 shows the demand displaced by a partial-storage system.

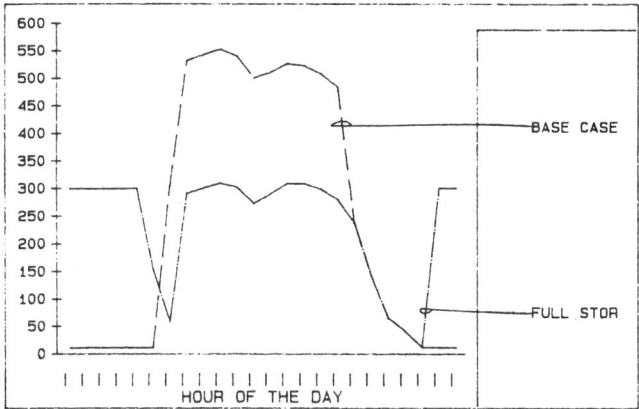

Figure 7-10. Peak kW demand profiles: base case versus full storage (San Diego)

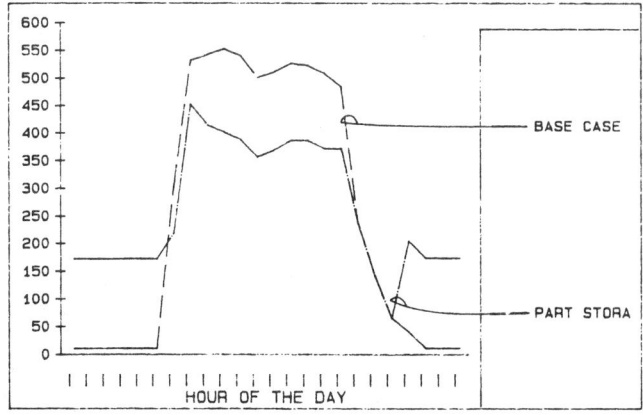

Figure 7-11. Peak kW demand profile: base case versus partial storage (San Diego)

Table 7-4 contains monthly cost savings delivered by both of these storage systems. Full storage enables the San Diego Gas and Electric customer to avoid $14,667 in demand charges and $30,237 in energy charges each year. Partial storage yields savings of $7,136 and $22,962 respectively. Note that in these cases, the avoided energy charges are greater than demand charge savings. This fact underscores the importance of having the capability to track electricity use hour-by-hour to accurately assess costs arising from the time-of-day rate structure. The San Diego Gas and Electric rate structure includes peak, intermediate, and off-peak energy charges as well as peak-period and intermediate-period demand charges. Hourly simulation with a sophisticated computer program gives the analyst the capability to accurately assess utility costs.

Capital incentives offered by San Diego Gas and Electric for kW peak avoidance systems are $225/kW of displaced cooling demand for retrofit application and $350/ton storage capacity for new construction. These cash subsidies amount to $54,675 for full storage and $22,500 for partial storage retrofit systems serving this building. These systems offer 3-year and 4-year simple paybacks.

If this were a new construction project, payback economics would benefit from larger utility subsidies amounting to $350 per ton of storage capacity for both full- and partial-storage systems. Subsidies for this building would total $120,050 for the full-storage option and $47,250 for partial storage. In addition, the chiller and cooling tower serving the partial-storage tank could be down-sized to realize another $26,709 in savings. The end result is a 2-year payback for full storage and a 3-year payback for the partial-storage system in San Diego.

CONCLUSIONS

Effective analysis of thermal storage systems requires consideration of many factors:
- The utility rate structure must be analyzed to determine whether it is conducive to thermal storage.
- The storage media must be selected to best fit the existing situation.

	Jan	Feb	Mar	Apr	May	Jun	Jul	Aug	Sep	Oct	Nov	Dec
Full Storage												
Peak kWh	$2591	2166	2589	2891	2803	2827	3455	3281	3053	2980	2584	2536
Intermed.	700	643	755	909	1248	1371	1673	1667	1475	1311	742	740
Off-Peak	−1005	−791	−982	−1319	−1514	−1792	−2351	−2187	−2054	−1734	−1152	−943
kW Demand	813	1155	1021	1254	1191	1495	1755	1399	1817	1475	1438	926
TOTAL	3099	3172	3383	3735	3728	3901	4532	4160	4291	4032	3612	3259
Partial Storage												
Peak kWh	$2590	2165	2588	2889	2801	2761	3001	3182	2844	2875	2525	2535
Intermed.	270	266	327	334	418	382	245	378	313	382	277	299
Off-Peak	−704	−583	−745	−1003	−1066	−1271	−1517	−1503	−1397	−1215	−889	−736
kW Demand	812	526	1021	395	362	463	646	322	708	436	519	926
TOTAL	2968	2374	3191	2615	2515	2335	2375	2379	2468	2478	2431	3024

Table 7-4. Monthly cost savings: San Diego

- The size of the storage tank can be selected to maximize savings or provide the best overall return.
- Secondary HVAC equipment can be down-sized if ice storage systems are considered.
- The yearly energy savings which can result from thermal storage are best analyzed by use of a computer program utilizing hourly calculations.

Thermal storage systems are becoming more and more attractive as utility companies increase their charges and adjust their rate structures. The best way to determine if thermal storage is right for your facility is to retain a competent engineering firm to analyze the problem for you.

References

Specialized Energy Study: Thermal Storage Systems. EMC Engineers, Inc., Technical Report. Denver, Colorado: 1986.

Thermal Storage. ASHRAE Technical Bulletin. Atlanta, Georgia: 1985.

"Thermal Storage Systems." *Heating/Piping/Air Conditioning,* Vol. 57, No. 1 (January 1985), pp. 133-151.

8

Case Study:
Three Proposed Cold Thermal Storage Systems for a U.S. Navy Shore Facility

J.F. McCauley

INTRODUCTION

About the time of Christ after the first Persian invasion of India a magnificent fortress, known today as the Red Fort, was constructed on the dry Delhi plateau. It was so large that cavalry mounted on elephants could engage in combat for sport on the fort's parade grounds.

In order to humidify and cool the ruler's quarters and the quarters of his harem, natural draft ventilation was directed through water sprays and into their living spaces. Here one of the earliest forms of air conditioning can be found and today is demonstrated for tourists.

During the era of the Revolution an inventor, Thomas Jefferson, introduced to America the practice of storing ice formed during the winter in semi-underground ice houses. It was used for cooling liquid refreshment and food preparation during the hot humid summers encountered in the central and southern portions of North America. Blacks were then in slave bondage and provided inexpensive labor to gather the ice. The idea caught on and most of those who could afford an ice storage house built one.

These perhaps were the first successful attempts at thermal storage known to man.

Today we can assemble sophisticated HVAC and thermal storage plants producing thousands of times more units of comfort than could our predecessors but not without an economic sacrifice. HVAC and thermal storage plants cost money. And one must choose from the various technologies which system meets the cost/benefit restraints for the project.

This chapter deals with an organization's experience in attempting to retrofit existing HVAC systems with accepted cold thermal storage technology equipment.

The HVAC equipment ranges in size from 25- to 550-ton chillers of the reciprocating, centrifugal and screw genuses. The types of storage systems studied are water, ice and eutectic salts; with single, dual and multi-tank storage arrangements.

Early in the study we determined water storage best suited our needs and we proceded to study the pros and cons of tank arrangements in deatil.

BACKGROUND

The Naval Aviation Supply Offices Scenario

Thermal storage is a concept aimed at reducing the electric utility costs associated with operating HVAC power chillers such as at the Naval Aviation Supply Offices facility in Philadelphia, Pennsylvania.

The capacity for temporary storage of thermal energy enables at least partial decoupling of the chiller from the building cooling load for large parts of the cooling period.

Arrangement of chiller operation during off-peak periods permits the activity to take best advantage of the provisions in the electric rate structure. Specifically, demands can be leveled off, thereby reducing demand charges. Also, usage during periods of high energy costs (peak periods) can be minimized. These savings opportunities are most significant when applied to buildings which currently are cooled by the electric chiller during the period of peak daytime rates and have few or no nighttime cooling requirements.

Three existing building systems at NASO exhibited characteristics inviting thermal storage study in conjunction with their HVAC systems.

Building No. 36, an administration office, is cooled by a 400-ton electric centrifugal chiller. The chiller load is primarily for comfort cooling and is typically required in service only 11 hours a day. It therefore was considered to be an ideal candidate.

Building No. 4, which comprises offices and computer areas, has a 360-ton electric centrifugal chiller. It also is characterized by loads primarily confined to the period from 6:00 a.m. to 5:00 p.m.

Finally, Buildings No. 2 and 3, which contain large computer areas and some offices, share a commonly connected chiller system comprising three large electric centrifugal chillers at 550 tons, 500 tons and 200 tons, and one electric reciprocating chiller with a capacity of 25 tons. Although Building No. 2, the Data Processing Center, contains computer equipment which requires a continuous cooling load, most of the load from Building No. 3 dissipates at night. Because this comfort cooling portion of the system load accounts for over 50% of the system capacity, it was considered a viable candidate for application of the thermal storage concept.

METHODOLOGY OF THE STUDY

An Activity Energy Plan study performed at NASO under the sponsorship of Northern Division, Naval Facilities Engineering Command, uncovered a need for a detailed study of air conditioners over 50 tons. Therefore, an Act-Up Phase I* Study was authorized for the Activity, with Northern Division overseeing the effort.

The Architect/Engineering firm selected to perform the study uncovered what seemed to be a natural opportunity to futher enhance several air conditioning systems at NASO. Although they did not do so formally, they stated that thermal storage and the cooling systems of several buildings appeared compatible. They pointed out the very large electrical operating cost savings potential of thermal storage systems.

---

*Act-Up: Air Conditioning Tune-Up Study. Act-Up Phase I studies cover air conditioning units over 50 tons capacity. Act-Up Phase II studies cover A/C units 50 tons or less.

Recognizing the Phase I study offered excellent energy conservation opportunities, the Activity has acted to cut energy costs. The study we are discussing, now known as a Specialized Energy Study (SES), is the outcome of the Act-Up Phase I study recommendation. The astuteness of the A/E performing the Act-Up I study and his deliberations with the Activity Public Works Officer were prerequisite to Northern Division's again soliciting proposals for the thermal storage study as described herein.

Bldg. No. 36 and proposed thermal storage tank.

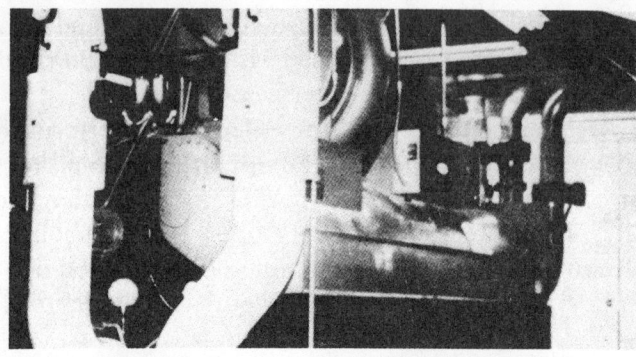

Bldg. No. 36, 400-ton centrifugal chiller.

The methods used to assess the technical and economic feasibility of the potential thermal storage systems are explained below.

Engineering Field Survey and Data Acquisition: Data was collected which relates to the operation and physical characteristics of existing HVAC equipment at NASO. This data and field investigation covers the following:

- Metered electrical consumption from the main base meter,
- Hourly demand profiles for typical days during the cooling season,
- Operating schedules (peak-shaving, daytime-energy-use reduction),
- Operation and equipment information for the centrifugal chillers,
- Drawings and layouts of the chilled water and HVAC systems,
- Building construction data such as wall and roof construction and area and window type and area,
- Occupancy and equipment-use schedules for existing buildings,
- Interior load data for occupancy, lighting and computer and office equipment loads and
- Tentative locations for the thermal storage systems.

Computer Analysis: The data collected during the field survey was analyzed by the DOE-2.1C computer program. The building envelope, building occupancy schedules and interior loads were simulated using actual Philadelphia weather. A building was first simulated as it currently exists and then simulated with the thermal storage system. The spare capacity of the chiller plant, which is available during off-peak hours, will be put into simulated use to supply chilled water to the thermal storage system. Again, the simulated chiller plant will be shut down during the peak cooling period and chilled water obtained from the thermal storage system to meet the building loads. It is expected the thermal storage plant will reduce the peak electrical demand and the electrical costs by operating the chillers during periods when utility rates are minimal.

Cost Estimate: The cost of installing thermal storage equipment was estimated for each system analyzed. The costs include storage tanks and their support structures, chilled water piping, pumps and controls.

Operating Schedules: The options in operating schedules hinge directly on the nature of electric rate schedules and building electric load profiles. With ratchet demand charges* and time-of-day rates (such as those at NASO), the scheduling objectives in thermal storage concepts are as follows:

- Peak Shaving: Levelized electric demand throughout the 24-hour day, thereby reducing peak kW charges and
- Daytime Energy-Use Reduction: Reduction of energy use during peak-rate daytime periods.

If chiller power requirements constitute over half the daily electric peak load and/or there are significant off-peak loads, the ideal storage operating schedule could call for some chiller operation during the day. This cooling could be provided in parallel with "storage tank discharging" to eliminate the chance of establishing a new peak at night.

At NASO, however, the ideal candidate buildings are characterized by steep profiles with nearly all air conditioning, lighting and miscellaneous electric loads currently disappearing at night (5:00 p.m. to 6:00 a.m.). Furthermore, power requirements for chilling the water constitute less than half the daytime total electric loads. Therefore, by rescheduling chiller operations as completely as possible to nighttime (off-peak) hours, the 24-hour electric peak can be reduced by the full amount of eliminated daytime chiller demand. Rescheduling can transfer a maximum amount of kWh usage from the high-cost daytime period to the low-cost nighttime period. This was the approach considered in evaluating the thermal storage concepts described below.

---

*Ratchet demand charges: see *"A Favorable Electrical Power Rate Structure"* - p. 4

THREE TECHNICAL CONCEPTS AND THEIR ECONOMICS

There are many ways to plan a thermal storage plant. In general, the approaches to implementing a thermal storage system involve combining the following components:

- Heat transfer equipment (centrifugal, reciprocating and/or screw-type chiller compressors),
- Storage media (water, ice, eutectic salts),
- Containment systems (single- or multi-container)

Heat Transfer Equipment: There are three generic types of refrigeration equipments readily available for use with a thermal storage system. Screw, reciprocating and centrifugal chillers are all capable of producing temperatures adequate for chilled water, ice or eutectic salt thermal storage systems. However, each also has limitations.

Each type is available with direct expansion (DX) evaporators. Also available are screw type and centrifugals which use flooded evaporators to evaporate the refrigerant. Direct expansion refrigeration systems are not recommended because they use thermal expansion valves and evaporator superheat in a delicate balance of variable suction and discharge conditions to get unevaporated refrigerant out of the suction gas. When this fails, the liquid remaining in the suction gas can be quite damaging to a reciprocating compressor. Consequently, the reliability of DX machines used in thermal storage applications is not high.

Screw machines (also called helical rotary machines) are a positive displacement type. They offer many attractive performance characteristics, including a nearly constant refrigerant flow rate, low noise, low speed, high thermal efficiency and low discharge temperature. This last point qualifies the screw-type chiller as an icemaker. However, the low discharge temperature comes at a high price. The 20 degree F suction temperature necessary to build ice on a coil requires about 30 percent more input energy than does the 42 degree F evaporator temperature that produces chilled water.

Reciprocating machines, too, are of the positive displacement type. Because pressure rise has only a slight influence on the volume flow rate of the compressor, a reciprocating chiller can maintain nearly full cooling capacity even on days when wet-bulb temperatures are above design conditions for the machine. This type of chiller is also capable of making ice. It can, however, be subject to the same energy penalties as the screw compressor in the ice-making temperature ranges. Another disadvantage of the reciprocating chiller is its capacity limit. The largest readily available machine of this kind is only about 200 tons. Because reciprocating chillers commonly use a DX evaporator, they do not work well in thermal storage applications. The varying suction and discharge conditions under which thermal storage systems operate can damage the compressor.

Centrifugal machines are the most widely used commercial HVAC chillers. Their popularity stems from their reliability, energy economy and ability to respond to fluctuating load conditions. Centrifugal chiller reliability is partially due to the use of flooded evaporators. These drop unevaporated refrigerant out of the suction gas, thus protecting the compressor. Because centrifugal machines are not of the constant displacement type, they offer a wide range of capacities continuously modulated over a limited range of pressure ratios. Thus, centrifugal chillers are desirable for both close temperature control and energy conservation because of their wide load handling characteristics (with nearly proportionate changes in power consumption).

However, centrifugal chillers face initial and operating costs when used for ice-making. These costs are not encountered when the chilled medium does not undergo a state (solid/liquid) change. A water/glycol solution is required as the heat transfer medium to make ice. The addition of the initial glycol cost and the cost of periodically replacing the glycol solution make ice storage systems driven by centrifugal chillers even more expensive than they already are.

Thermal storage for a new construction project presents options that cannot be considered for retrofit applications such as those at NASO. In new construction, the heat transfer equipment can be selected with thermal storage in mind. For instance, in new construction a screw or centrifugal chiller compressor could be sized exactly for the load and choice of storage medium.

In retrofit construction, as is the case at NASO, the designer must use existing chillers unless those units are old and ready for replacement. At NASO, the chillers serving Buildings No. 2, 3, 4 and 36 are not ready for replacement. But by using storage system options of chilled water or eutectic salts, NASO could take full economic advantage of the existing equipment and would not have to install new heat transfer media or glycol. The following section further discusses important issues involving the various types of storage media.

Storage Media: Water is the most common storage medium. It provides a capacity of 1 btu per lb-degree F in the liquid state and 144 btu per lb when changing from liquid to ice at 32 degrees F. Whether the additional thermal storage capacity associated with the phase change offers a net benefit depends largely on the specific application.

The efficient production of ice may require refrigeration equipment different from conventional HVAC chillers. When properly selected, ice-making systems can operate at efficiencies comparable to HVAC chiller systems. But with retrofit HVAC applications such as those at NASO, the ice-making approach would add to the initial cost of the project, unlike using existing equipment to chill water for storage in the liquid state.

Containment Systems

Schemes for containing chiller water in thermal storage systems focus primarily on balancing vessel volumes and costs with the need to keep warm (unchilled) water separate from chilled water. (The separation is necessary to ensure that the chillers operate at favorable evaporator temperatures during the off-peak hours.) Three different approaches are commonly taken:

Single Tank: The first approach utilizes a single tank with no excess volume but with a flexible bladder dividing the enclosed space (see Figure 8-1). With this approach, the volume of water in the tank remains constant while the relative amounts of warm and chilled water vary. During the night (off-peak period), warm water from the upper storage volume is pumped through the chiller and into the lower part of the tank. At 6:00 a.m. the lower tank (cooled water) is lined up to discharge through the cooling coils and back into the upper tank to be chilled during the night. The cycle repeats itself.

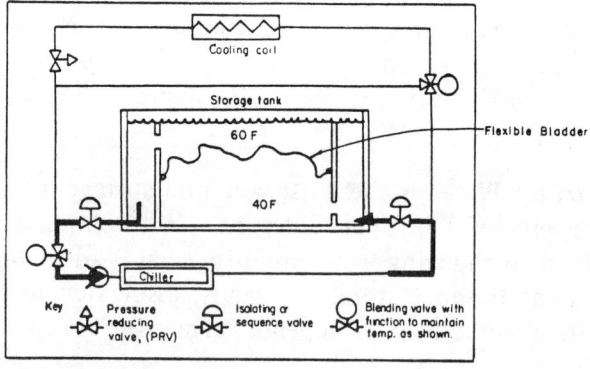

a) Tank charging mode

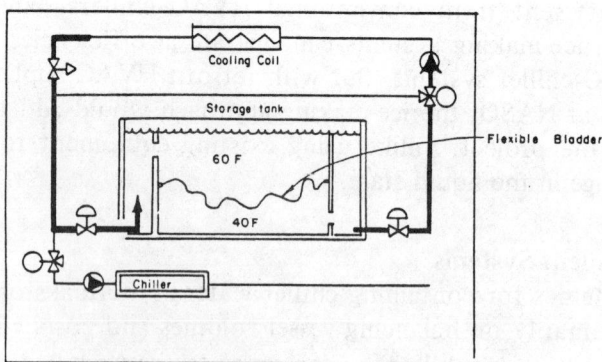

b) Tank discharge mode

Figure 8-1. Chilled water storage, single tank with flexible bladder.

Dual Tank: The second approach utilizes two tanks, each sized to contain the full volume of the chilled water to be generated during off-peak hours. During the day (on-peak period), one tank is filled with return (warm) water while the other tank is emptied of chilled water to satisfy loads. At night (off-peak period), the warm water is pumped from the full tank, through the chillers, and into the empty tank in preparation for the cooling loads of the next day. Problems with this approach include the hazard of air being drawn into the system as either tank empties. Also, this approach requires installation of a total tank volume which is twice that needed to contain the system water at a given time. This may not be possible in retrofit projects where space for tank installation is limited.

Multiple Tank: The third approach would partially alleviate the requirement for excessive storage volume. A larger number of tanks, perhaps six, might be used, each with a capacity one-fifth of the total system volume. See Figure 8-2. During the off-peak nighttime hours, warm water from the tanks flows through the chiller and into the empty storage volume. When cooling needs occur the following day, chilled water is delivered to the cooling coils. The warm water discharged from the coils fills the available storage volume over the course of the day, and at night the system repeats the cycle. In this way, separation between warm and chilled water can be maintained with only 20% excess storage volume. The hazard of air being drawn into the system remains, however, and the cost and complexity of tank construction and piping assembly increase.

The single tank approach was selected for NASO to minimize first costs and space requirements for tank installation.

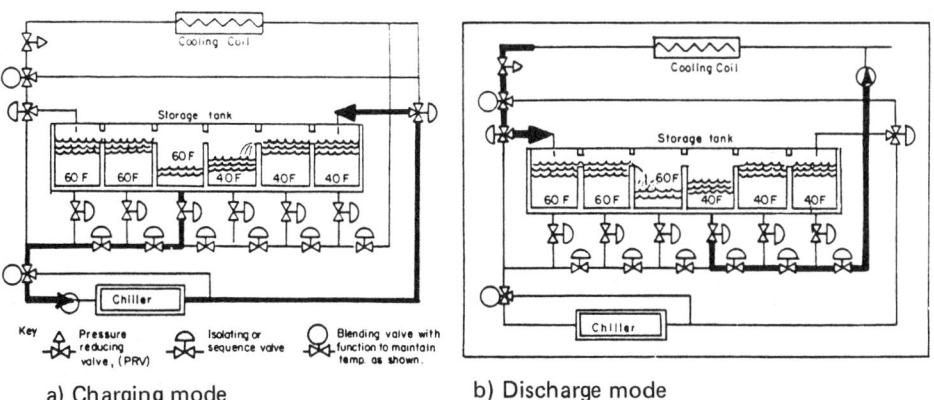

Figure 8-2. Chilled water storage, multiple tank system.

A FAVORABLE ELECTRICAL POWER RATE STRUCTURE

The electric company power rate structure may include part or all of the following elements:

- Time-of-day rate adjustment,
- Seasonal rate adjustment,
- kWh energy rate,
- kW demand charge,
- kW demand ratchet clause and
- Customer charge.

NASO is provided electric service by the Philadelphia Electric Company under their high tension power rate. This rate comprises charges and credits for electric energy use according to a time-of-day and seasonal structure. On-peak hours are defined as 8:00 a.m. to 8:00 p.m., Monday through Thursday. Friday on-peak hours are from 8:00 a.m. to 4:00 p.m. All other hours are off-peak, including weekends and holidays.

The time-of-day energy use is but one component of the monthly electric bill. A customer service charge and peak power demand constitute the remaining charges. Kilowatt-hours are billed at three incremental rates. The first 150 hours times the kW demand is charged at $0.0739/kWh. The second 150 hours times the demand is billed at $0.0556/kWh. The balance of the monthly kWh usage costs is $0.0376/kWh. Each kilowatt of monthly peak demand is billed at $5.37/kW. The customer charge is $220.45 per month.

Kilowatt-hours consumed during on-peak hours receive the additional on-peak charge of $0.0057/kWh from June through September. The off-peak credit for kilowatt-hours used in the off-peak period during these months is $0.0021/kWh. The months of October through May have an on-peak and off-peak charge and credit of $0.0022/kWh and $0.0021/kWh, respectively.

The peak kW demand is billed according to the season. The billing demand for each month of the October-through-May period is 80% of the highest demand recorded during the June-through-

September period. Billing demand for each month of the June-through-September period is the actual peak kW occurring during the month. Table 8-1, following, summarizes the rate schedule.

Table 8-1. Electric Rate Schedule

Billing Category	Charges (in $)				Credit
	Per Mo	Per kW	Per kWh	Per on-peak kWh	Per off-peak kWh
Customer service	$220.45				
Demand		$5.37			
Energy					
1st 150 hrs x demand			$0.0739		
2nd 150 hrs x demand			0.0556		
Additional Wage			0.0376		
June through September (Summer Months)				$0.0057	$0.0021
October through May (Winter Months)				0.0022	0.0021

NORTHERN DIVISION'S TECHNICAL CHOICE

Buildings No. 2, 3, 4 and 36 were analyzed both with and without thermal storage. For brevity, however, only Building No. 36 analysis will be presented since what applies to Building No. 36 applies to all.

Existing Systems: Building No. 36 is currently served by a variable air volume system which is run in the cooling mode from May through September. The chilled water coils are fed by a 400-ton centrifugal chiller operating from 6:00 a.m. to 5:00 p.m. and providing cooling to the 117,000 sq. ft. space dedicated to general office-type use. The building is occupied from 7:00 a.m. until 6:00 p.m. weekdays. Because the chiller and fans are shut down at night, this schedule is well suited to using the unoccupied hours to generate and store chilled water for the cooling needs of the next day.

The daytime cooling loads are caused by various factors, signigicantly heat generated by lighting, office equipment and personnel. Other loads include: conduction through the building envelope, solar gains through windows and heat introduced with outside air through infiltration and ventilating air. Another source of heat is the gain from building mechanical equipment such as air conditioning, refrigeration and air handling equipment and domestic water heaters.

Occupant density is approximately 85 square feet per person, the lighting level is 1.65 watts per square foot and office equipment uses about 0.75 watts per square foot. Because of the total heat load of the building structure and its contents, heat is stored in walls and furniture during the day. When the mechanical system fans are shut down at about 5:00 p.m., ventilation is eliminated. Because heat stored in the building structure and contents is released into the building during the night, temperatures inside the building remain relatively warm despite cooling outside temperatures. Early morning solar gains contribute to the interior heat level. Consequently, cooling is sometimes needed at 6:00 a.m. or 7:00 a.m. to bring the interior temperatures of the building down to comfort range preparatory to employee arrivals. Even if cooling is not required before the workday starts, loads are rapidly generated as people arrive and lights and equipment are energized.

Figure 8-3 represents the hourly cooling load profile on a typical summer day. Building No. 36 was still warm when the chiller came on-line at 6:30 a.m. to cool the building in anticipation of the work day. The daily peak occurred during the second hour of chiller operation due to the increased cooling load of the work force, lights, office equipment and solar gains.

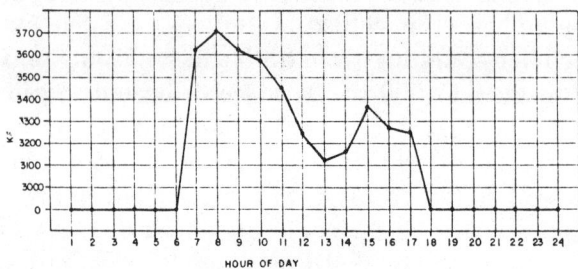

Figure 8-3. Bldg. No. 36, hourly cooling load profile — typical summer day.

Case Study: Proposed Cold Thermal Storage Systems for U.S. Navy Facility

As the cooling system caught up with heat gains, the cooling load dropped steadily throughout the remaining morning. Greatly reduced gains occurred when equipment was turned off and people left the building at midday. As a result, the cooling load bottomed at 1:00 p.m. and then climbed as afternoon activity increased, peaking at 3:00 p.m.

Figure 8-5 depicts the hourly electric demand profile for Building No. 36 on a typical summer day. The base electric load consists primarily of nighttime lighting. Demand took a jump at 6:00 a.m. when the 400-ton chiller switched on to begin cooling the building. Demand climbed steeply when lights and equipment were turned on to accommodate the workday. A morning peak was established at 10:00 a.m. when office activities were in full swing. The profile registered a dip as workers turned off machinery and left the building for lunch. Electric demand again increased as people returned to work. The peak demand of the day occurred at 3:00 p.m.

Proposed Storage System: The chilled water storage system investigated in the study employs the existing 400-ton chiller to generate chilled water during the 5:00 p.m. to 8:00 a.m. off-peak electric period. This system produces 800 gallons of water per minute cooled an average 10 degrees F in one pass through the chiller.

An insulated storage tank located above the ground and capable of holding 705,000 gallons is divided into upper and lower sections by a movable membrane. While the tank is charging, the chiller receives 50 degrees F to 60 degrees F water from the upper section and supplies the lower section with 40 degrees F to 50 degrees F chilled water. The membrane moves up inside the tank as the upper water volume diminishes and the lower section volume increases.

During the day, cooling loads demand chilled water from the lower tank section and circulate it through the cooling coils of the Variable Air Volume (VAV) air handlers. It is returned to the upper section of the storage tank 10 degrees F to 14 degrees F warmer than when it left the lower section. As the volume of the lower section is depleted and the volume of the upper section grows, the membrane moves down inside the tank. At the end of the working day, the volume of chilled water has been drawn off and is again ready for the tank charging mode of operation. The cycle repeats daily, with the charge mode beginning Sunday evening at 8:00 p.m. and ending Friday with the shutdown of building operations. The system sits idle over the weekend until the charge mode begins again on Sunday in anticipation of the workweek. This daily cycle is in use from May 1 through September 30.

Figures 8-1 and 8-4, respectively, illustrate the tank membrane system and the proposed configuration of the storage system.

Figure 8-5 depicts the hourly electric peak demand profile for Building No. 36 with and without a storage system, on a typical summer day.

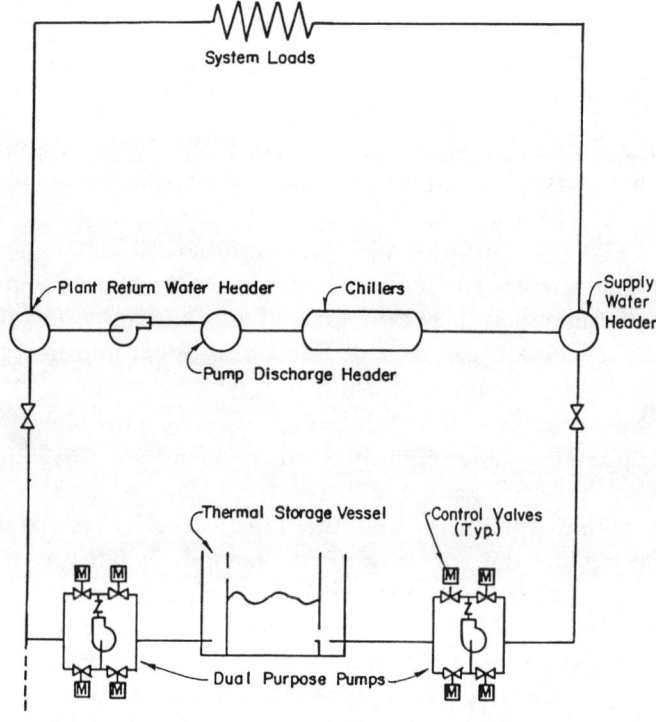

Figure 8-4. Proposed thermal storage system.

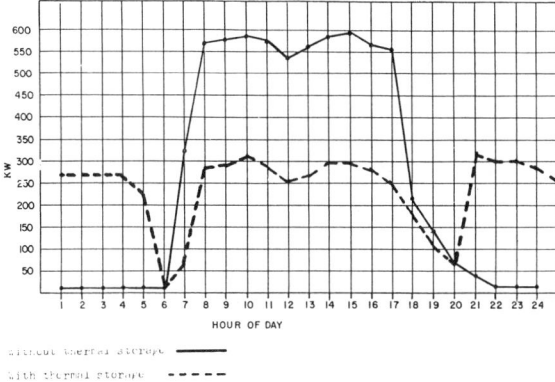

Figure 8-5. Bldg. No. 36, peak electric demand profile with and without thermal storage, typical summer day.

Summary of Energy Cost Savings: Cost savings could be realized by shifting chiller operation from on-peak electric hours to off-peak, taking advantage of the lower off-peak electric rates. Additional cost savings could be obtained by reducing kilowatt demand during the period that the facility-wide peak occurs. This peak occurs during the day when lighting and equipment demand are concurrent with peak cooling loads. Shifting the operation of the chiller in Building No. 36 would reduce the building peak power demand by the amount of power drawn by the chiller (up to 271 kilowatts). Consequently, both the facility-wide daytime demand and the kilowatt demand charge would drop correspondingly.

Tables 8-2 and 8-3 summarize the base case and storage case electrical usage respectively and present a summary of electrical data. The storage system reduces on-peak usage by 166,172 kilowatt hours, the total electrical usage by 13,646 kilowatt hours and the building peak kilowatt demand from 593 to 332 kilowatts.

Table 8-4 displays the electric cost savings delivered by the chilled water storage system. Savings in the non-cooling season are the result of the 80% ratchet clause in the utility rate structure.

Table 8-2. Building No. 36, cooling season electrical loads, base case summary

Month	Peak kW	On-peak kWh	Off-peak kWh	Total kWh
May	524	86,121	22,930	109,024
Jun	854	91,249	29,815	121,064
Jul	590	105,446	25,919	131,365
Aug	593	98,022	26,889	124,911
Sep	587	90,062	22,798	112,860
Total	593	470,900	128,351	599,224

Table 8-3. Building No. 36, cooling season electrical loads, storage case summary

Month	Peak kW	On-peak kWh	Off-peak kWh	Total kWh
May	311	59,140	42,029	101,169
Jun	322	60,677	58,966	119,643
Jul	332	64,650	66,563	131,213
Aug	322	60,179	64,536	124,715
Sep	318	60,082	48,756	108,838
Total	332	304,728	280,850	585,578

Reducing the cooling season peak demand by 261 kilowatts causes the billing demand during the noncooling season to drop by 80% to 209 kilowatts. The total cost savings realized by the storage system are $37,894 per year.

Storage System Cost Estimate: The cost of the proposed chilled water system is estimated at $450,000, based on 1985 material and labor costs. This value was derived by using standard engineering cost estimating methods. Individual component costs were taken from "MEANS 1985 Building Construction Costs Data" or were quoted by equipment suppliers. Table 8-5 summarizes component costs.

Case Study: Proposed Cold Thermal Storage Systems for U.S. Navy Facility 161

Table 8-4. Bldg. No. 36, electrical cost savings with thermal storage.

Month	Savings (1985$)
Jan	2,993
Feb	2,993
Mar	2,993
Apr	2,993
May	3,392
Jun	4,055
Jul	4,037
Aug	4,201
Sep	4,251
Oct	- 0 -
Nov	2,993
Dec	2,993
Total	37,894

Table 8-5. Bldg. No. 36, thermal storage machinery cost estimate

		COST ESTIMATE			DATE PREPARED 11-13-85		SHEET 1 OF 1	
ACTIVITY AND LOCATION BLDG. 36, CHW STG. CONCEPT W/SINGLE TANK & BLADDER			CONSTRUCTION CONTRACT NO.				IDENTIFICATION NUMBER	
			ESTIMATED BY GT				CATEGORY CODE NUMBER	
PROJECT TITLE NAVFAC, NORTHDIV NASO, PHILADELPHIA, PA			STATUS OF DESIGN PED ☐ 30% ☐ 100% ☐ FINAL ☒ Other (Specify) Feas Stdy				JOB ORDER NUMBER	
ITEM DESCRIPTION	QUANTITY		MATERIAL COST		LABOR COST		ENGINEERING ESTIMATE	
	NUMBER	UNIT	UNIT COST	TOTAL	UNIT COST	TOTAL	UNIT COST	TOTAL
STEEL TANK	1	EA					LS	146,000
FOUNDATION							LS	46,200
BLADDER							LS	50,000
TANK INSULATION	9,500	SF	.09	855	.25	2,375		3,230
PUMPS	2	EA	5,100	10,200	1,500	3,000		13,200
INSULATED PIPING	500	LF	48.15	24,075	21.60	10,400		34,875
VALVES/FITTINGS							LS	7,000
ELECTRIC POWER WIRING							LS	1,750
CONTROLS							LS	5,000
SUBTOTAL								307,255
OP & L (27.05%)								83,112
CONTINGENCY (15%)								58,555
TOTAL								448,922
ROUNDED TO								450,000

BASIS FOR DECISION ON
EXERCISING A PROJECT DEVELOPMENT OPTION

Thermal storage systems were investigated as economical means of reducing the cost of space cooling at the Naval Aviation Supply Office. Buildings No. 2, 3, 4 and 36 were studied as prospective sites of storage systems. The preceding are detailed discussions of benefits to be derived from Building No. 36 HVAC modification with a thermal storage system. It is typical of the other three buildings with respect to thermal storage.

Chilled water was selected as the storage medium because of its lower first cost in retrofit applications compared to other available storage media. The Philadelphia Electric Company provides power to NASO through the High Tension Rate. This rate structure is conducive to the nighttime generation of chilled water which can be stored and used to address the cooling loads of the following day. The chilled water approach allows the use of existing chillers to charge storage tanks.

The costs of installed components for ice and eutectic salt systems for Building No. 36 are higher (28 and 36%, respectively) than the chilled water storage system cost estimated for this building. Savings-to-investment ratios and simple payback periods will therefore be less attractive for these systems than for a chilled water system.

The results of this study indicate chilled water storage delivers an annual dollar savings on electric costs for each building and building complex of $37,894, $44,464 and $65,515 respectively for Buildings No. 36; 4; 2 and 3.

The electric cost savings resulting from installing a thermal storage system are substantial. However, the costs of retrofitting buildings with chilled water storage systems are also substantial. Chilled water storage systems exhibit economies of scale. The installed costs of the 705,000-gallon systems studied at Building No. 36 is $0.64 per gallon. Tanks of larger size, however, start to require additional structural considerations which increase the costs disproportionately to the 705,000-gallon tank.

Another factor influencing storage system economics is the amount of time during the year the system operates. The cooling season for Building No. 36 is only 5 months. During the remaining 7 months, the system sits idle and generates no electric cost savings. In contrast, the system serving Buildings No. 2 and 3 is operated year-round, producing cost savings every month.

The higher system cost per gallon of storage, combined with 7 months of idle time each year, spells relatively long payback periods for chilled water storage systems serving Building No. 36. The system serving Buildings No. 2 and 3, on the other hand, enjoys a shorter payback as a result of a system economy of scale and year-round usage.

The Navy uses the Savings/Investment Ratio (SIR) as their economic guide in qualifying projects for funding. Energy Conservation Projects with SIRs greater than 2.5 are considered valid from the viewpoint of taxpayer dollars equitably spent in shore facility projects. Table 8-6 highlights the techno/economics of the three projects.

Table 8-6. Summary of storage system savings.

Building	On-peak kWh Saved	Peak Demand Saved	Dollars Saved	Savings-to-Investment Ratio (SIR)
2 & 3	1,085,944	399	65,515	1.02
4	188,819	313	44,464	0.80
36	166,172	271	37,894	0.94

Conceding electrical costs could increase and possibly economically justify thermal storage, one must realize that there is a straight line relationship between electrical costs and the SIR in this project. Therefore, to attain an SIR of 2.5 (other factors being equal), electrical costs would have to increase by a factor of almost 1.6 for the Building No. 36 project to qualify economically.

As well as qualifying economically, the project must also qualify technically. A legitimate need for the Energy Conservation Project must be demonstrated also. In the above case a technically viable, legitimate need has been demonstrated. But the Economic Analysis indicates the Government can better invest capital elsewhere.

Some of us feel we could substitute underground wells as a holding (storage) place for water at 45 to 55 degrees F. When wells are drawn on, air handler chillers would only be required to cool the water 5 to 10 degrees F. The solution in not as simple as it seems. It involves environmental, forensic and electro/mechanical barriers of formidable magnitude. However, storage wells open a door for us when thermal storage closes a door to further investigation.

Summarizing our efforts in this thermal storage study, we find that capital investment in retrofitting Navy equipment at NASO is not compatible with the minimum savings expected of an Energy Conservation Project.

The industrial community, however, may consider an investment in a cold thermal storage plant like this because of various tax incentives it enjoys, which of course are unavailable to government. The tax credit for investing in new machinery and the tax credit for energy conservation investment, combined, could render this or similar cold thermal storage schemes very attractive. My intention in preparing this chapter was not to express how a good idea can fall in the acid environment of dollars and cents realism but rather to offer our experience with a project's failure. Hopefully, industrial practitioners of engineering, working within less rigid economic restraints, will have more success.

SUMMARY
Methodology of the Study
A discussion on Northern Division's procedure in studying the feasibility of using thermal storage in conjunction with three large HVAC chillers at the Naval Aviation Supply Office, Philadelphia, PA, was executed.

Technical Concepts
We have discussed the science of three cold thermal storage systems and the relative merit of each from the view of capital investment, maintenance and electrical cost saving.

A Favorable Electrical Power Rate Structure

The necessity for an electric company to offer a rate structure permitting low cost operation of refrigeration equipment during off-peak hours was scrutinized.

Northern Division's Technical Choice

A discussion of the pros and cons of the technology and economics of the three thermal storage system sciences was undertaken. They were considered as they relate to NASO chiller plants. Attention was given to the choice of retaining or replacing the existing chillers as each thermal storage system may demand.

Basis for Decision on Exercising a Project Option

The project's bottom line, after assuring the technical feasibility of thermal storage at NASO, was the Saving-Investment Ratio. This is used throughout the Navy Shore facility complex as a "go-no go" tool for determining energy saving project viability.

ACKNOWLEDGEMENTS

Appreciation is expressed to Lieutenant Commander J.L. Donofrio, Public Works Officer, along with Robert Bloch, P.E. and the able staff of engineers and their support personnel at the Naval Aviation Supply Office, Philadelphia, Pa.

Also thank you, Bruce Bechard, P.E. and the staff of EMC Engineers, Inc., contracted to perform the thermal storage study, for producing an outstanding report upon which the author relied in preparing the preceeding chapter.

Special thanks is extended to my colleagues and managers whose capable critique rendered the chapter worthy of presentation at the World Energy Engineers Congress, 1986.

And also, thank you, Mr. Dick Koral, AEE for your encouragement to undertake the writing of this chapter.

9

Case Study:
Thermal Energy Storage for Municipal Buildings

D.S. Teji, J.R. Balon

INTRODUCTION

Over the last 25 years, electricity has become the energy source of choice for the heating and cooling of large commercial buildings. A major side effect of this practice has been a steady escalation in the size and sharpness of utility demand peaks. Most buildings tend to experience their peak cooling load at a similar hour of the afternoon, and their peak heating load at a similar hour of the morning. This coincident demand for electricity taxes the capacity of the utility to keep up, and can even lead to "brownouts," temporary shortages of electricity. The utilities have begun to react by offering rates and incentives which encourage night time or "off-peak" use of electricity in place of daytime or "on-peak" use. By doing so, the utilities hope to increase the load factor of their existing power plants, and postpone the burden of constructing new, much more expensive plants.

Conventionally heated and cooled office buildings have limited potential for taking advantages of off-peak rates, except for the possibility of thermal energy storage. This concept involves the use of electricity during the hours when it is cheapest, to charge a hot or cold thermal reservoir. The reservoir is then used to heat or cool the building during the hours when electricity would be most expensive to use. In this way, thermal storage reduces peak electrical demand, and takes advantage of utility rate incentives, without sacrificing the comfort of building occupants.

In 1985 the City of Phoenix carried out a 1-year applied research program in thermal storage for its buildings. The work was divided into three phases:
- Phase One – Background Research
- Phase Two – Building Survey
- Phase Three – Demonstration

This chapter summarizes the major results of each phase.

PHASE ONE – BACKGROUND RESEARCH

The purpose of this phase was to discover all information about thermal storage presently available. In fact information is still scanty, as thermal storage has just emerged from its infancy as a modern air-conditioning alternative. The utilities themselves have taken primary responsibility for generating both qualitative and technical information on thermal storage and dispersing it to the public. To date, the most comprehensive source of information is the Electric Power Research Institute (EPRI) *Commercial Cool Storage Design Guide* and related publications [1, 2, 3, 4]. (EPRI is the research organization representing electrical utility companies throughout the U.S.) Individual utilities which promote thermal storage also prepare technical manuals and hold seminars for the engineering and development communities. In 1984 the major centers of thermal storage activity were Chicago, San Diego and Dallas. They were joined by Phoenix in 1985 when both local utilities here announced new and major thermal storage incentive programs.

Technology Assessment

Reliability: Storage systems installed in the 1970's tended to be trouble prone, with problems stemming primarily from operator inexperience, faulty time clocks and manual controls, compressor abuse and failure, leaking storage vessels and inaccurate ice thickness controls. By 1985 nearly all of these problems have been eliminated or minimized. Most of the progress can be attributed to standardization and packaging of complete storage systems, using more appropriate refrigeration equipment and better engineering.

In new construction at least, the technology of cool storage is ready for use.

Retrofit Application: Thermal storage technology is more difficult to apply in retrofit situations. Existing refrigeration compressors are often not suitable for making ice, the preferred medium for retrofit applications. The major alternative medium, chilled water, requires about five times as much space as ice and locating room for it is problematic. Newer technologies which ease retrofit problems include *brine ice-banks* [5] which can be used with some existing reciprocating chillers to freeze ice at 32 deg F. *Eutectic salts storage* [6] holds additional promise for easing retrofit problems:

- It is compact because it stores cooling in the form of a phase change, like ice, but
- The phase change temperature is 43 deg F, rather than 32 deg F, which permits it to be used with any existing air-conditioning chiller.

Thus, cool storage technology is beginning to seriously address the retrofit market.

Operating Strategies

Most sources name the choice of control strategies for thermal storage as "full storage" or "partial storage," and sometimes "demand-limiting storage" is also mentioned. Most sources attempt to define these terms by illustrating how the strategy might be implemented for a particular utility rate for a particular building under theoretical design day conditions. The definitions are consequently *ad hoc,* "for this case only." In reality, there is no simple definition of each strategy which will apply to all situations. The meaning of the terms subtly shift, depending on the building load profile, the utility rate, and even the time of year.

To illustrate the effect of the utility rate on the control strategy, consider the following comparison. Figure 9-1 illustrates "full storage" under a *time-of-use* utility rate. The "off-peak" hours in this instance have been defined as 7 p.m. to 11 a.m. daily. During these hours the refrigeration compressors are allowed to operate at full load, to charge storage or to meet the building load directly. During the remaining, on-peak hours the compressors are shut down. Thus the compressors use only the cheaper, off-peak energy to cool the building all day.

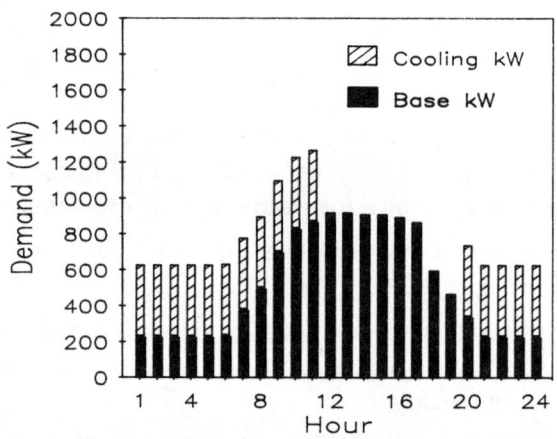

Figure 9-1. Full storage under a time-of-use rate.

In contrast, Figure 9-2 shows a "full storage" strategy for the same building under a *conventional* utility rate. Under a conventional rate there are no off-peak hours. The objective, accordingly, is only to limit the peak demand experienced by the building. Thus, the refrigeration compressors are controlled so as not to operate during hours of high demand in the building.

Comparing Figures 9-1 and 9-2 demonstrates that the structure of the utility rate indeed has a major influence on defining "full storage." In this example, the utility rate affects the amount of storage required, the size of the refrigeration plant, the hours of compressor operation and the nature of the controls. Moreover the picture is becoming more complex than this simple example suggests. Time-of-use rates often have "wrinkles" which further complicate the picture, for example:

- "Shoulder Rates"; an intermediate period between full on-peak and full off-peak rates,

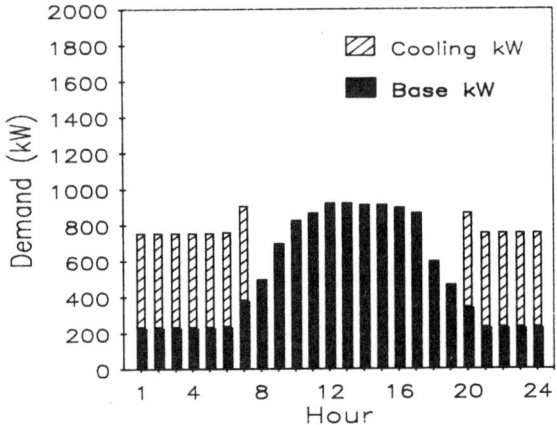

Figure 9-2. Full storage under a conventional rate.

- "Off-Peak Weekends" where energy charges are lowered all day Saturday and Sunday,
- "Sliding Demand Windows" which make the on-peak *demand* hours different from the on-peak *energy* hours,
- "Ratchet Clauses" which set a minimum monthly demand based on the maximum annual demand, and
- "Seasonal Rates" where the rate charges and rate structure may change from summer to winter.

Each of these features may affect "optimal" control of the storage system. But in addition to accommodating these complexities of utility rate structures, the control strategy may change with changing loads on the building. For example, a system designed for partial storage at design conditions may in fact operate as a full storage system for most of the year when the cooling load is not so high. Most sources gloss over these complexities and, at present, no comprehensive treatment of control strategies exists. Consequently, some trial and error must still be expected in developing a control strategy.

PHASE TWO – SURVEY OF CITY BUILDINGS

Table 9-1 summarizes results of the cool storage survey of city buildings. The first column gives the building name and the second gives the estimated storage capacity required for demand-limiting storage. The unit of storage capacity is the "ton-hour," which is the cooling produced by one ton of refrigeration operating for one hour (12,000 Btu's of heat removal). Storage sizes range from 3800 ton-hours for the Municipal Building (city hall) down to 70 ton-hours for a small fire station. The third column holds the estimated cost of installing the required storage. This cost includes adjustments for the scale of the projects; the cost per ton-hour declines with increasing size. The cost also reflects how extensively the existing HVAC system must be modified; existing direct expansion (DX) systems must be converted to chilled water distribution during a thermal storage retrofit. The fourth column lists the expected incentive payment from the utility, based on $250 per kW shifted off-peak. The fifth column shows the annual savings calculated for the building after storage has been installed and a time-of-use rate is in effect. The last column shows the payback for the retrofit, after allowing for a 6% annual rate of escalation in electricity prices.

The paybacks range from 3 to 12 years with a roughly even distribution between these extremes. Before the utilities announced their incentive program for thermal storage, the payback on storage retrofits was 17 years or more. The incentives and time-of-use rates have turned this situation around so that paybacks are now 12 years or less. In the most favorable cases the paybacks are in the range of 3 to 6 years.

The type of existing HVAC equipment at a building strongly influences the payback. All storage systems at present require a chilled water distribution system. If the existing distribution system at the building is not chilled water then the system must be converted to chilled water. Such a conversion can be moderately to prohibitively expensive, depending on the details of the existing air-distribution system. Among the surveyed buildings the most costly retrofits would be those with packaged DX air-conditioners, carrying a conversion cost of $150 to $220 per ton-hour. Compare this cost with a

Table 9-1. Results of cool storage survey of city buildings.

Building Name	Storage Size (Ton-Hrs)	Retrofit Cost	Incentive Payment	Annual Savings	Payback (Years)
Police Bldg.	2720	$255,000	$102,000	$45,000	3
Municipal Bldg.	3800	$360,000	$ 93,000	$42,000	6
Plaza Municipal	2400	$230,000	$ 74,500	$22,000	6
Central Library	2000	$225,000	$ 70,000	$25,000	5
Art Museum	500	$ 56,000	$ 16,250	$ 3,150	10
Police Academy	500	$ 64,000	$ 13,750	$ 4,100	9
LEAP No. 3	300	$ 48,000	$ 16,500	$ 3,500	8
Field Engineering	500	$ 75,000	$ 21,250	$ 3,800	10
Adult Center	300	$ 22,000	$ 14,750	$ 2,000	4
Little Theater	500	$ 64,000	$ 20,000	$ 3,000	11
Peuble Grande Museum	200	$ 35,000	$ 9,500	$ 1,500	12
Police Briefing Sta.	300	$ 38,000	$ 9,000	$ 2,600	9
Fire Academy	150	$ 24,000	$ 9,000	$ 1,700	8
Central Library	200	$ 16,000	$ 7,500	$ 4,500	5
Fire Support Service	100	$ 22,000	$ 4,500	$ 1,200	10
Fire Station No. 21	70	$ 12,000	$ 2,000	$ 600	12

range of $73 to $128 per ton-hour for buildings with existing chilled water systems. The type of existing HVAC system can make a difference of a factor of two in the cost of a storage retrofit.

Even in buildings with chilled water distribution, complications can arise. Storage must be located at or below grade because of its weight, requiring that space be located for it around the outside of the building. Ideally the refrigeration equipment room will also be located on the first floor or basement of the building. Otherwise, interfacing the storage with the refrigeration compressor may be expensive or impractical.

PHASE THREE — DEMONSTRATION

System Selection

Ice was selected as the storage medium for the following reasons:
- Space was at a premium in all the buildings which were being considered for a retrofit.
- Chilled water storage tanks above ground would be unsightly while burying would require extensive tearing up of facilities.

- Ice banks are available as manufactured units with known cost. The cost of chilled water storage, on the other hand, depends heavily on site-specific factors and would not be known until late in the design process. This delay would complicate budgeting and planning the project.

Among ice builders one must further choose between *brine-type* and direct expansion (*DX*) type. Both types build ice on the outside of a "coil" of pipes or tubes by circulating a coolant on the inside of the tube. The DX type ice builders use refrigerant directly as the coolant. Thus, in a DX ice builder the ice-holding pipe coils are physically the evaporator of the refrigeration cycle. A brine ice builder, on the other hand, uses an anti-freeze mixture of water and glycol as the coolant inside the tube coil. The glycol mixture of "brine" is cooled to subfreezing temperature by a standard, packaged air-conditioning or refrigeration chiller. For the demonstration project we selected brine-type ice banks, with the following reasons:

- The brine ice bank could make use of the existing reciprocating chiller at the test site.
- Using a secondary coolant isolates the chilled water loop from the storage water, preventing contamination.
- Brine ice builders behave reliably. They produce even "build up" and "burn off" of the ice in storage, avoiding the problem of "tunneling" and "bridging" common in DX ice builders.

Overall, brine ice builders provided the quickest and least expensive avenue to use for this particular demonstration project.

Building Selection

The building selected to receive a retrofit was a branch library with approximately 6000 square feet of conditioned area and a peak cooling load of 18 tons. Century Library was selected because it was small enough to be retrofit within the budget of $20,000, because its existing chiller and chilled water distribution system could be used with minimum modification, and because it had a representative load profile for city buildings.

Peak electrical demand at the library was 55 kW, over half of which (30 kW) came from air-conditioning equipment. The annual electric bill, prior to the retrofit, was $10,522, with approximately $3,300 going toward cooling.

Electric Rate Description

Table 9-2 shows the existing, conventional electric rate on which previous year's charges were based. The conventional rate features a declining block rate structure. This structure lowers the rate for electricity as usage increases, irrespective of the time of day and the actual cost to produce the electricity. An expanding first block and seasonally adjusted rates are also incorporated.

Table 9-2. Existing Electric Rate at Century Library

Customer Type:	General Commercial Class	
Energy Charge:	Summer	Winter
First 4,000 kWh	$0.0747/kWh	$0.0615/kWh
Next (75 x Peak kW)	$0.0747/kWh	$0.0615/kWh
Next 50,000 kWh	$0.0472/kWh	$0.0437/kWh
All Additonal Kwh	$0.0357/kWh	$0.0300/kWh
Demand Charge:	$2.78/kW	$1.55/kW

Table 9-3 shows the optional time-of-use rate to which the building was converted. Notice that the declining block structure is gone, replaced by a division of rates according to the time of day. This type of structure more accurately reflects the utility's cost of producing electricity. To further reflect its costs the utility has chosen to retain its summer/winter rate split, and includes different hours as "on-peak" in the different seasons. Weekends are off-peak in the winter but not in the summer.

Selection of the Storage Size

The cooling load calculations for Century Library indicated a total daily cooling load of 203 ton-hours. Of this total, about 70 percent or 140 ton-hours would occur during on-peak hours (12

Table 9-3. Optional Time-of-Use Rate.

Customer Type:	Experimental, Thermal Storage	
On-Peak hours:	Summer — 12 noon to 10 p.m. DAILY	
	Winter — 7 a.m. to 10 p.m. M - F	
Energy Charge:	Summer	Winter
All On-Peak kWh	$0.0743/kWh	$0.0528/kWh
All Off-Peak kWh	$0.0294/kWh	$0.0294/kWh
Demand Charge:		
On-Peak Demand	$4.15/kW	$2.84/kW
Off-Peak Demand	No Charge	No Charge

noon to 10 p.m.) Thus, full storage could theoretically be accomplished with 140 ton-hours of storage capacity. However, an allowance of 10% was made for losses from the storage tanks and piping, bringing the requirement up to 154 ton-hours. Next, discrete tank sizes had to be selected. The CALMAC storage tanks selected for this project come in nominal sizes of 60, 90 and 100 ton-hours. However, the two smaller sizes are designed for quick charging and discharging and were not appropriate for this application. Two of the 100 ton-hour tanks were therefore necessary to cover the anticipated load. On the surface then, storage rated at 200 ton-hours was installed to meet a cooling load of 140 ton-hours, which appears to be an overdesign of 42 percent. In reality, the 100 ton-hour tanks could effectively supply 90 ton-hours at full load, because of limitations in heat transfer. Thus, an available 180 ton-hours (2 x 90) was installed to meet an anticipated cooling load plus losses of 154 ton-hours, producing a 17% "safety factor." Most engineers currently design storage systems with a safety factor of 15-25%. Thus the factor of 17% used in the design was not unreasonable.

System Piping and Control

Figures 9-3 and 9-4 illustrate the final design for the ice storage retrofit of Century Library. Figure 9-3 shows the charging cycle for the ice bank. The charging cycle begins at 10 p.m. and continues through the night until storage is fully charged. In this cycle the ice bank receives coolant from the chiller at 26 deg F, uses the coolant to freeze ice at 32 deg F, and discharges the coolant at 32 deg F. Next, an automatic diverting valve diverts the flow of coolant from its normal path through the building and sends it directly back to the chiller. This process continues until the ice bank is frozen solid and latent heat transfer ends. When latent heat transfer is complete the discharge temperature from the ice bank will begin to fall below 32 deg F. When the discharge temperature falls to 29 deg F a thermostat shuts down the chiller and the coolant pump and ends the charging cycle.

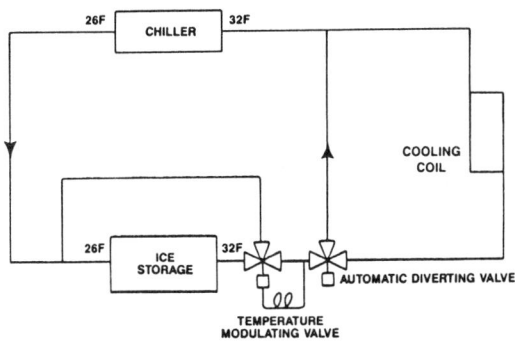

Figure 9-3. Charging cycle.

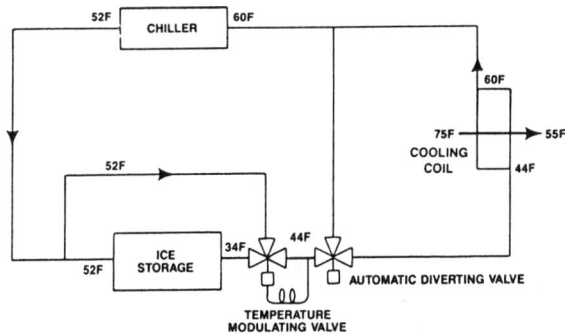

Figure 9-4. Discharging cycle (partial storage mode).

During the next day (8 a.m. to 10 p.m.) the system discharges the storage to cool the building. As shown in Figure 9-4, the automatic diverting valve switches flow back to the cooling coils of the building. The temperature modulating valve mixes 34 deg F coolant from the ice storage with return coolant from the building to produce 44 deg F coolant for the cooling coils in the building. At this time the chiller may be fully on, fully off or part-loaded, depending on the control strategy being used. Figure 9-4 shows the case in which the chiller is allowed to operate at partial load and supply part of the cooling load simultaneously with the ice bank (partial storage). If the chiller were now turned off the 60 deg F return temperature from the building would go on to the ice bank without a reduction in temperature. The motorized temperature modulating valve would react and proportion the flow so that the ice bank picks up the entire building load (full storage). Finally, if the chiller is allowed to operate at full load it will chill the coolant down to 44 deg F by itself. The modulating valve will sense no need for supplemental cooling and will bypass the ice bank (conventional operation). This system design has the following advantages in retrofit situations:

- Full storage, partial storage and conventional operation can be activated as desired from this single design.
- The existing chiller can be used to make ice and to cool the building directly when required by the control strategy.
- The existing chilled water distribution system, consisting of cooling coils, 3-way valves, air-distribution fans and circulating pump, can be used without modification because the temperature and flow rate of the coolant is the same before and after the retrofit.

The master control for the storage system was a Honeywell "W7000" energy management system, already being used at the building. Output channels of the EMS microprocessor were wired to relays controlling the chiller, motorized valves, thermostats, air-distribution van and coolant pump in the ice-storage system. The channels could then be programmed with the on/off times dictated by a given control strategy. The EMS microprocessor has a self-contained battery backup which will maintain the correct time and program information in the event of a power interruption. Good storage system design required a power backup feature on all time controls. Thus, linking the storage system to the existing EMS system for the building was a convenient and reliable method of controlling the storage system.

Instrumentation and Monitoring

The installation was fully instrumented to monitor the thermal and electrical performance of the ice storage system. An IBM-PC microcomputer was used to conduct datalogging from the various instruments which included wattmeters, flowmeters and thermocouples. The PC read data from the instruments continuously and recorded average values to disk every 15 minutes, 24 hours a day. The data was recorded in a LOTUS 1-2-3 compatible format for later analysis using this spreadsheet software.

The system was monitored for 6 weeks of severe summer weather conditions. During the course of the experiment the system was operated in all three possible modes; full storage, partial storage and conventional operation. The findings are presented below.

Results

Figures 9-5 and 9-6 portray a typical day during which the system was monitored. Figure 9-5 shows the ambient air temperature recorded on site at the library. Figure 9-6 shows the hourly cooling load on the building, recorded on the same day. The cooling load is nearly flat, a feature characteristic of buildings which are not cooled at night, and which have heavy construction and little window area. The load peaks near 10 a.m. because of the "pull down" in temperature which must be accomplished after air-conditioning resumes in the morning.

Conventional Operation: Figures 9-7 and 9-8 profile the electrical consumption of the building under conventional chiller operation. Figure 9-7 shows the total electrical demand while Figure 9-8 isolates the energy used by chiller. Compare Figures 9-6 and 9-8 and observe how the chiller electrical demand is tied to the building cooling load. Under conventional operation the chiller must follow the load almost exactly. Since the bulk of the air-conditioning load occurs during peak hours (12 noon to 10 p.m.) so does the bulk of the chiller energy consumption. In fact the chiller consumed 75 percent of its daily energy during "on-peak" hours.

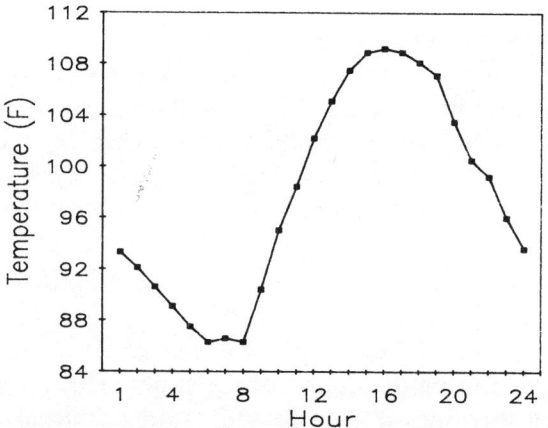

Figure 9-5. Ambient temperature.

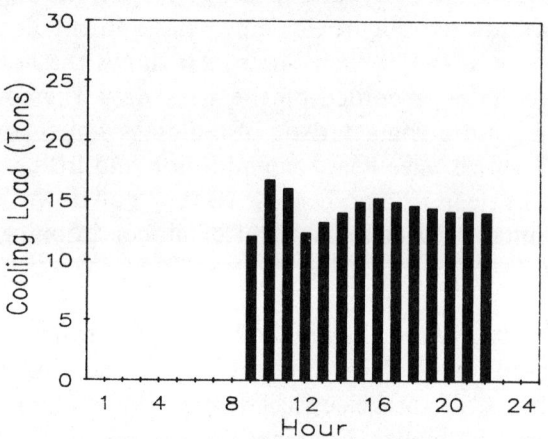

Figure 9-6. Building cooling load.

Case Study: Thermal Energy Storage for Municipal Buildings 181

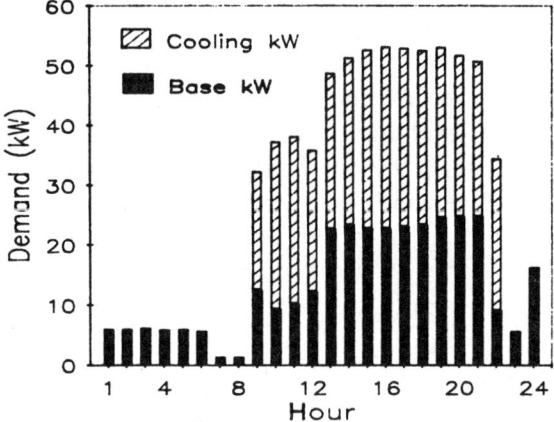

Figure 9-7. Building kW under conventional operation.

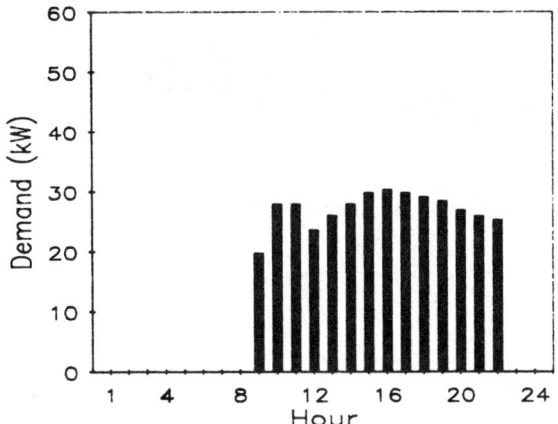

Figure 9-8. Cooling kW under conventional operation.

Partial Storage Operation: Under partial storage operation the chiller was allowed to operate at part-load during the daytime. The chiller was then capable of picking up about half the cooling load from building directly, with the ice bank making up the difference. At night the chiller continued to work at part-load to recharge the ice bank. Figures 9-9 and 9-10 illustrate the effect of the partial storage strategy on the building energy profile. The effect is to "level out" both the chiller energy use (Figure 9-10) and the building energy use (Figure 9-9-) over 24 hours. Hence partial storage is sometimes also called "load levelling." Partial storage reduced the on-peak consumption of energy by the chiller from 75 percent down to 38 percent. The peak demand level was reduced from 55 kW to 45 kW.

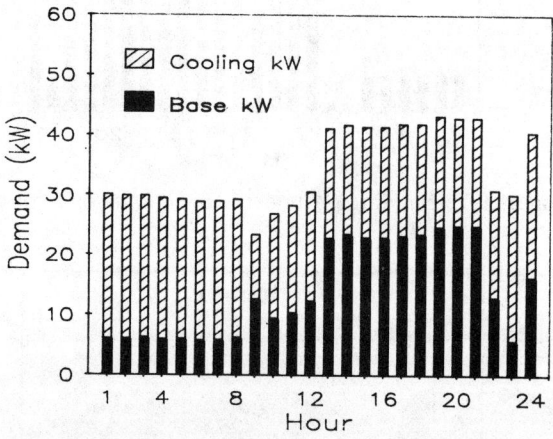

Figure 9-9. Building kW under partial storage.

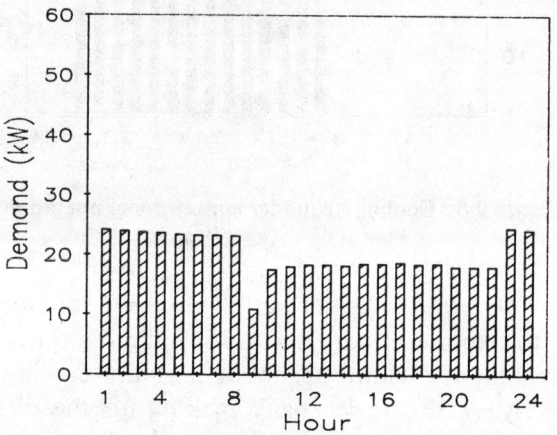

Figure 9-10. Cooling kW under partial storage.

Full Storage Operation: Under the full storage strategy the chiller was not allowed to operate during the peak hours of 12 noon to 10 p.m. daily. The ice storage alone served the cooling needs of the building during these hours. During the hours of 8 a.m. to 12 noon the building required cooling, but off-peak rates were still in effect. There would be no point in cooling the building from storage during off-peak hours, so the chiller was allowed to operate as a conventional system for these 4 hours and meet the cooling load directly. During the remaining 10 hours (from 10 p.m. to 8 a.m.) the chiller operated at full capacity to recharge the ice storage banks.

Figures 9-11 and 9-12 portray full storage operation. Full storage successfully removed all chiller energy consumption from peak hours. The recorded peak demand has been reduced by over half, from 55 kW down to 25 kW. The fact that off-peak demand reaches 59 kW does not matter because off-peak demand does not count under the time-of-use rate. Overall, only 28% of the total building energy consumption remains on peak.

Changes in Chiller Efficiency: The average efficiency of the chiller changed only slightly between conventional operation, full storage and partial storage. The specific energy use (kW/Ton) of the chiller rose by 3% under full storage and by 7% under partial storage. Most of the increased energy use under partial storage was due to the inefficiency of part-load operation for the existing chiller.

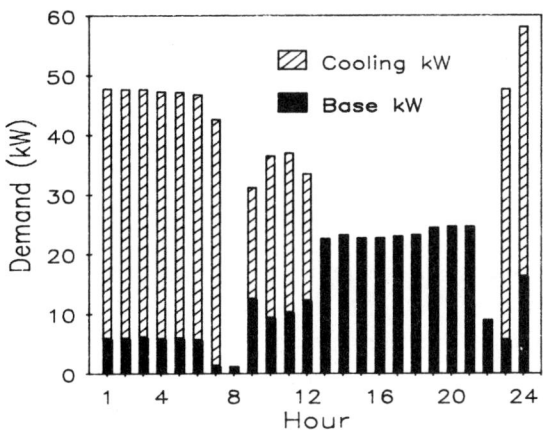

Figure 9-11. Building kW under full storage.

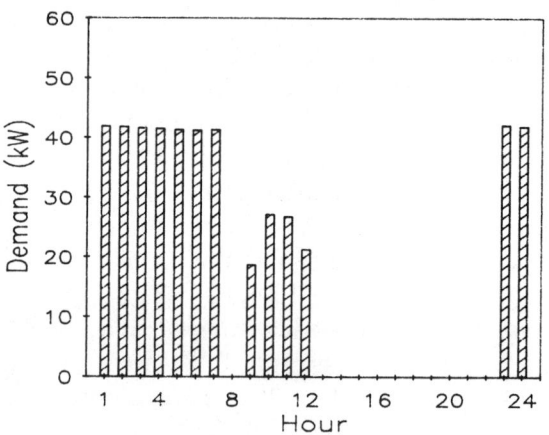

Figure 9-12. Cooling kW under full storage.

Changes in Total Cooling Energy with Storage: The total energy used for cooling per day rose 23 % under full storage and 28% under partial storage. Contributing factors consisted of the decrease in chiller efficiency described above, thermal losses from the ice banks and piping, and additional run time on the circulating pump and condensing fans. The increased energy consumption was included in calculation of annual cooling costs and deducted from the system savings.

Annual Savings and Paybacks: Comparing first cost to annual savings, the full storage system would pay for itself over a 5-year period:

Total Cost in Installation	− $16,000
Utility Rebate	− $ 7,500
Annual Operating Savings	− $ 1,650
Payback Period (16000-7500)/1650	− 5 years

Both the time-of-use rate and cash rebate were necessary to make the payback period attractive. Without the time-of-use rate the savings in operating cost would have been almost negligible. The payback period would have been completely unacceptable, on the order of 45 years. With the time-of-use rate the savings jumps to over half the annual cost of air-conditioning. But even with these impressive savings the high first-cost of storage would have made the payback period 10 years. The incentive payment from the utility was needed to bring the payback period down to 5 years, an acceptable time for us.

CONCLUSION
Technology Assessment
- Thermal Energy Storage technology is ready for widespread application, but is not yet completely mature.
- Thermal Storage savings are significant. Savings of 40-60% in the operating costs of air-conditioning systems are possible.
- Strong utility incentives, including time-of-day rates and rebates are needed to make thermal storage attractive.
- Retrofit options are often limited because of
  - Incompatibility with existing equipment
  - Lack of space for storage
- The number of types of systems and equipment vendors is increasing rapidly.

Suggestions for Application
The following procedure is offered for consideration as a possible path to use in investigating thermal storage opportunities:

1) *Consult Your Local Utility* company at the outset for information on thermal storage incentives. The utility may offer case incentives, time-of-use rates, preliminary feasibility studies and so on. If the utility has no incentive program, thermal storage will not be economical (at least for retrofits) in your service area.

2) *Rank Buildings by Preliminary Feasibility.* Create a list of buildings ranked by preliminary feasibility. First rank the buildings by the type of existing HVAC system:

    First Priority — Chilled Water Systems.
    Second Priority — Central DX Systems.
    Third Priority — Multiple, Packaged DX.

Within each category, further rank the buildings by the *location* of the existing equipment and the *availability* of space for storage. The ideal situation will have the existing equipment room located near ground level with space for storage immediately adjacent, either on-grade or under a lawn or parking surface.

3) *Roughly Estimate The Cooling Load* over on-peak hours. An estimate accurate to 25% can be obtained from examining the operating log of the refrigeration equipment or by analyzing the monthly electric bills for the building.

4) *Estimate Paybacks.* At this point, obtain rough payback figures for one or two buildings from each of the priority categories. Use rules of thumb for the unit cost of storage and the annual operating savings. This step will indicate which categories are attractive.

5) *Do a Full Engineering Study* on buildings with attractive paybacks. Compare the basic system alternatives; chilled water, ice and eutectic salts, on feasibility and payback.

6) *Obtain Competitive Bids* on the final design.

7) *Install The System.* If possible, install and test the system in the off-season for cooling.

ACKNOWLEDGEMENTS

Funding for this project was provided by the Energy Task Force of the Urban Consortium for Technology Initiatives under a grant from the federal Department of Energy. A copy of the complete, 120-page report on the project may be obtained for a small charge. Contact:

>Publications and Distribution
>Public Technology, Inc.
>1301 Pennsylvania Avenue, NW
>Washington, DC 20004
>
>Phone (202) 626-2443

Request document number DG/85-307 02/86-100: *Thermal Storage Strategies for Energy Cost Reduction.*

References

[1] EPRI EM-3371: *Commercial Cool Storage Primer.* Electric Power Research Institute, P.O. Box 50490, Palo Alto, CA 94303. January 1985. To order, phone (415) 965-4081.

[2] EPRI EM-3981: *Commercial Cool Storage Design Guide.* May 1985. Phone (415) 965-4081.

[3] EPRI EM-4125: *Current Trends in Commercial Cool Storage.* July 1985. Phone (415) 965-4081.

[4] EPRI EM-4405. *Commercial Cool Storage Presentation Material.* Volume I: Seminar Handbook. February 1986. Phone (415) 965-4081.

[5] CALMAC Manufacturing Corporation, Englewood, NJ 07631-0710. Phone (201) 569-0420.

[6] TRANSPHASE SYSTEMS INC. Thermal Energy Storage Systems. Huntington Beach, CA. Phone (714) 841-4010.

SECTION IV
EVAPORATIVE COOLING

10
Survey of the Indirect Evaporative Cooling Field

J.R. Watt

INTRODUCTION

Many companies now offer indirect evaporative cooling equipment acceptable in all but a handful of mostly coastal U.S. counties.

These cool air in heat exchangers separate from the evaporating water, so deliver dry *supply* air which can be further cooled by direct evaporative or refrigerative second stages. Power savings of 50% over all-refrigerative systems are common, plus first, maintenance, and winter heat-recovery economies.

THE OTHER EVAPORATIVE COOLING

Direct evaporative coolers like drip and other wet-pad coolers and some air washers, cool air *adiabatically* in wet porous pads which evaporate water into it. This removes no heat; instead, at constant *wet-bulb temperature* it converts the air's *sensible heat* into *latent heat* in the absorbed water vapor. This lowers the air's *dry-bulb temperature* perhaps 70-90% of original *wet-bulb depression* (dry-bulb minus wet-bulb temperature), so creates a damp cooling medium usually called *washed air.*

Its humidity prevents its recirculation indoors like refrigerated *supply air,* so it must be exhausted outside after one pass through the cooled premises. Much of its cooling effect depends upon contacting human skin at velocities above 100 fpm (30.5 m/min), so large volumes and high room-entering velocities are needed. Power savings over refrigeration around 70% result, but with those handicaps, direct evaporative cooling has little use outside arid regions except where massive ventilation is needed.

INDIRECT EVAPORATIVE COOLING

Indirect evaporative cooling is more like refrigerative air conditioning; it occurs at constant *dew-point temperature,* and removes sensible heat without adding humidity. The air being dry-cooled, termed *primary air,* is processed in finned-coil, tubular, plate-type, rotary, or other heat exchangers which separate it from the *secondary air* into which the water evaporates and which carries away the unwanted humidity.

The primary air is usually cooled 40-70% of outside wet-bulb depression, so is cool enough for good comfort-cooling supply air only in the driest weather. Further, the air flow resistance of most heat exchangers precludes using the volumes and circulating velocities of direct cooling. Thus, almost all indirect coolers have second stages of either direct cooling or small-scale refrigeration to lower the cooled-air temperatures further for better comfort effect.

In the former case, washed air is created, subject to discharge outdoors after brief use; this wastes perhaps 50% of its cooling potential, but the overall process still saves at least half the power of refrigerated cooling. In the second case, the cool air remains dry and becomes supply air comparable to that from all refrigerated systems. It can be recirculated as return air as in refrigerated cooling, and also used *regeneratively* (see below).

Because both types of second stage effectively extend the wet-bulb depression, staged indirect cooling operates far outside its original Southwest, the former type in all but a few, mostly sea coast areas of Alabama, Arkansas, Florida, Georgia, Louisiana, Mississippi, South Carolina, and Texas.

Systems with refrigerated second stages, of course, operate anywhere. Because the latter is relieved of most sensible cooling, power savings over total refrigeration remain significant, ranging around 25%.

Regenerative gains are commonly sought, especially in buildings with large ventilating needs. To remove indoor pollutants, these must exhaust large volumes of used cooling air. If this is still essentially dry, adding it to entering secondary air lowers the latter's wet-bulb temperature, creating cooler primary air and improving the economy.

Naturally, washed air from indirect-direct two-stage systems cannot be so used. However, such air routed into the air-cooled condensers of refrigerated air conditioning or refrigerated second stages of other systems accomplishes a like result.

Most indirect systems also save winter energy for such buildings by pre-heating the new *make-up air* they draw in for ventilation. Here, warm used, indoor air being discharged becomes dry secondary air and conducts heat to fresh cold air entering the primary-air system. In cooler climates this heat-recovery saving may exceed that in summer.

AVAILABLE INDIRECT COOLING SYSTEMS

At least six indirect systems employ types of cooling towers and finned coils or equivalent dry surfaces. Three use plastic-tube heat exchangers, two with absorbent wick-like coatings. At least four firms make plate-type indirect coolers, two with thin plastic plates, two with absorbent coatings. Two other manufacturers use heat-transfer wheels, and one uses a desiccant-enhanced indirect system employing solar heat. All will be discussed in order.

Tower-and-Coil (Dry Surface) Systems.

These originated in Arizona before 1930, using natural-draft spray or deck towers and indoor auto radiators and fans. Oversized towers cooled water near 3 deg F (1.7 deg C) of wet-bulb, and the thick multi-pass finned water coils which replaced the radiators often cooled primary air equally near tower water temperature. In the hottest weather, supply air was often 30 deg F (16.7 deg C) below outdoor dry-bulb.

Unhappily, with scant tower bleed-off and water treatment, circulating hard water augmented by trapped dust lined the coil tubes with performance-destroying scale. Yearly acid removal deteriorated the coils and WWII copper shortages prevented replacement. The development around 1935 of plate-type indirect coolers and drip coolers delivering larger volumes of cooler air at a fraction of the cost also helped scrap tower-and-coil systems.

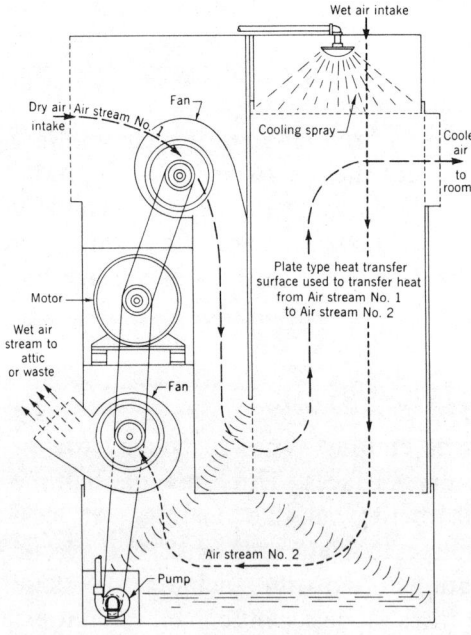

Figure 10-1. An early plate-type indirect cooler. Spreading from California, these cooled scores of Southwestern chain stores in the 1930s and 1940s.

With better bleed-off and water treatment they are reviving today, mostly in soft water areas:

First: *Norsaire Systems,* of Englewood, CO, which also makes large plate-type and rotary heat-exchanger wheel systems, uses powered, counterflow vertical towers and multi-pass coils followed by direct evaporative second stages for loads over 90 tons (1,139 mJ/hr).

Unknown engineers and contractors are following suit, best using "coil-shed" towers or heat-exchangers to separate tower and circulating water.

Survey of the Indirect Evaporative Cooling Field 195

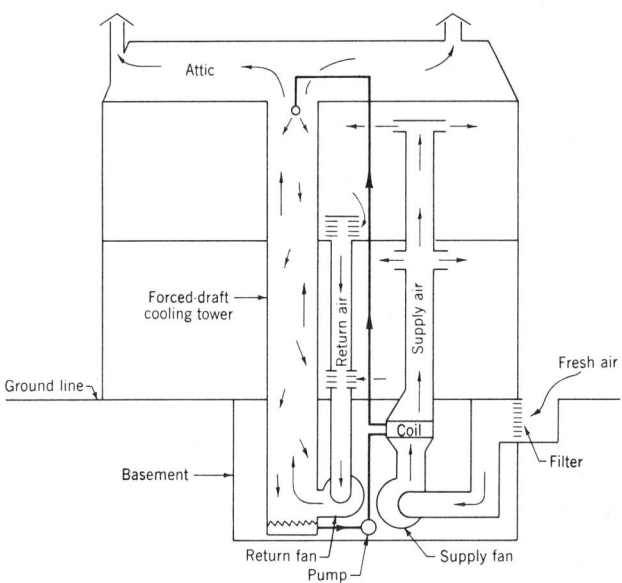

Figure 10-2. The University of Arizona administration building cooling system, 1936-1952. This demonstrated the best tower-and-coil cooling, regenerative use of return air, and attic removal of solar heat load.

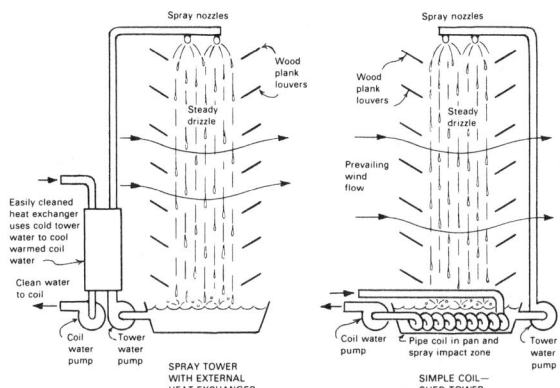

Figure 10-3. Cooling towers arranged to keep scale out of cooling coils. Unfortunately, they require extra pumps and cut performance slightly.

Second: "open loop" use of cooling towers in large chilled water systems is growing each winter. Here, buildings with large indoor human, lighting, and process heat gains all year stop their compressors when outdoor dry-bulbs approach 40 F (4.4 C), and circulate carefully filtered, bled-off and treated tower water through their chilled water coils and air handlers.

Cooled by atmospheric contact as well as by evaporation, such water deposits little scale, but some systems use heat exchangers or coil-shed towers for further protection. Reduced water flows in colder weather prevent tower icing. Power savings vary with weather and usage. Around New York, installation payback is reportedly 2 years.

Third: in some areas with large *diurnal* temperature swings between day and night, cooling towers are operated nightly at off-peak power rates to store cold water for next day use. In Colorado a college and a public school keep respectively 60,000- and 100,000-gal (227 and 379 kl) tanks below 58 F (14.4 C) for water-coil use.

As noted, such cool-air use of towers minimizes scaling tendencies. However, the large insulated tanks push first costs above those for refrigeration, but power savings suggest 4-year paybacks.

Fourth: *Aztech International, Ltd.,* Albuquerque, NM, manufactures very compact four-stage indirect-direct coolers containing two identifiable tower-and-coil sets plus multiple regeneration.

Stage 1: filtered outside primary air is cooled dry in metal tubes chilled by washed air from thick, high cooling-efficiency rigid-media saturation pads that serve as both direct coolers and towers. Stage 2: the primary air is next cooled in finned coils carrying water cooled in these pads. Stage 3: the same air is dry-cooled again in similar finned coils cooled by water from the *fourth* stage pads. Because three indirect stages have lowered primary-air wet-bulb temperatures, this water is regeneratively cold. Stage 4: the thrice dry-cooled air is adiabatically converted in a final high-efficiency wet pad to cold washed air, well below original wet-bulb temperature.

Made in several sizes to 8,000 cfm (3.8 m^3/s), these units should operate in almost any climate. However, water quality and ample bleed-off seem essential.

Survey of the Indirect Evaporative Cooling Field

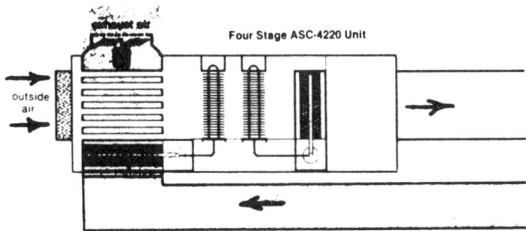

Figure 10-4. An Aztec four-stage indirect-direct cooler. Shown in winter heat-recovery use, it has two tower-and-coil pairs. Here, with pads and coils dry, used ventilating air in bottom duct warms fresh make-up air, left, in the Stage 1 tubular heat-exchanger, left center, and discharges, top.

Fifth: an historic system seems worth improving. In the 1940s, the late Professor B.N. Gafford of Electrical Engineering, University of Texas at Austin, built a compact regenerative rooftop single-stage cooler. It consisted of two large vertical truck radiators connected so a 1/6 hp pump circulated sealed-in water between them.

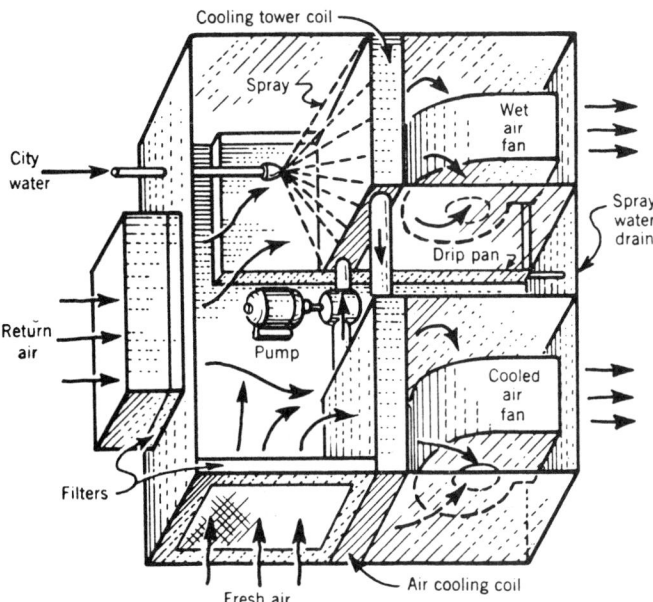

Figure 10-5. The 1940s Gafford regenerative single-stage cooler. The cooling-tower coil, rear, sprayed with city water, and the closed-circuit water circulation prevented internal scaling. The used water flooded a flat roof against solar heat.

One, serving as coil-shed tower, had city water mist sprayed on its fins and discarded while mixed outdoor and return house-air was drawn through it. Water cooled inside it circulated through the other radiator and cooled primary air.

When the author tested it in 1953, it still cooled 3,711 cfm (1.75 m³/s) of air within 10 deg F (5.6 deg C) of outside wet-bulb for about 5/6 hp. Maintenance costs and deterioration seemed nil. Use seems recommended where the waste water can be profitably used.

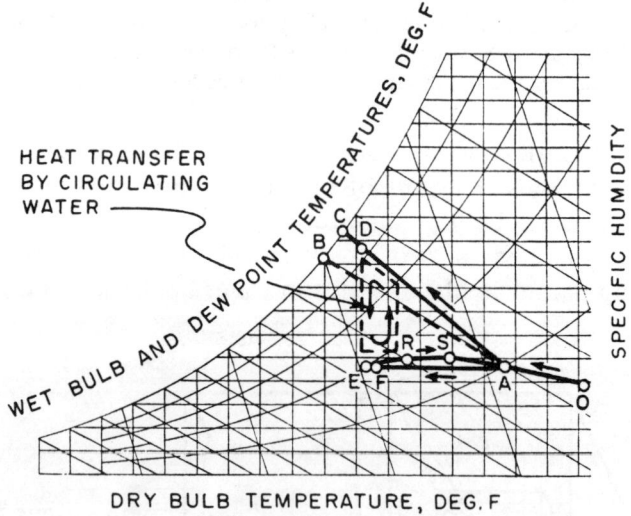

Psychrometric Process in Regenerative Dry-Surface and Gafford Indirect Evaporative Coolers. Outside air O and return air S mix, forming air A of wet-bulb temperature B. Nonadiabatic process AD cools water to DE, which cools dry air from A to E. It enters room at F and absorbs heat to S past room average R. Heat loss AE equals heat gain AD. (Not to scale.)

Figure 10-6.

Finally: *Howden Heat Pipe Division,* Bloomfield, CT, offers a very new but similar system composed of one or more vertical grids of closely spaced horizontal, sealed and finned *heat pipes*. A partition divides the grids into two panels. The tube halves in one panel serve as coil-shed cooling towers; water sprays above and in front wet them as secondary air is drawn around and between them. Fresh primary air blows through the other panel.

Survey of the Indirect Evaporative Cooling Field

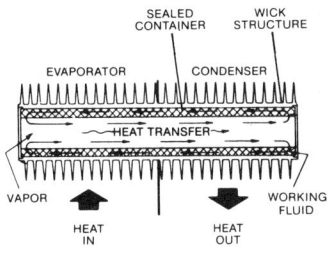

Figure 10-7. Inside a Howden heat pipe. Heat at either end evaporates liquid refrigerant and sends vapors to condense at the other, while wicking returns the condensate to the warm end. Continuous end-to-end heat transfer results in even 20 ft (6.1 m) lengths.

The tubes are lined with wick-like absorbent and contain measured amounts of appropriate refrigerant gas. The wet cooling-tower ends condense it to liquid which moves by capillary action to the air-cooling ends, where it absorbs primary-air sensible heat through the metal walls. Vaporized, it returns as gas to the cool end, where it condenses again and delivers the absorbed heat to evaporating water.

Thus, the heat pipes continuously transfer heat from primary to secondary air. They form seemingly foolproof heat exchangers only inches thick and with purely external and removable scale, if any. They serve as well for winter heat-recovery.

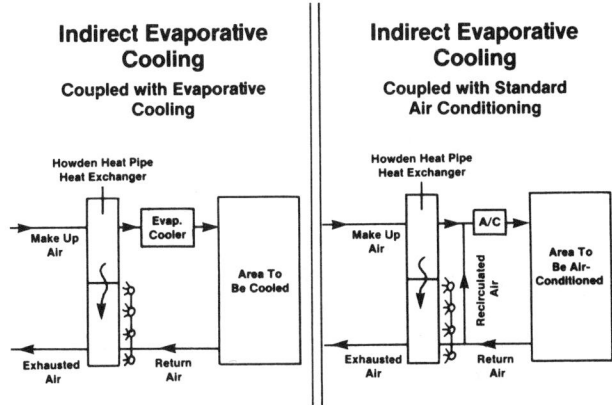

Figure 10-8. Staged Howden heat-pipe cooling. Left: with a direct second stage. Right, with a refrigerative one. The heat pipes conduct heat downward in each drawing.

These indirect coolers are available in many sizes up to 28 tubes 20 ft (6.1 m) long by 8 rows thick. Including fan and pump power they reportedly achieve Energy Efficiency Ratios (Btus removed per watt) as follows:

Single stage indirect cooler	119 EER
Two-stage indirect-direct cooler adjusted for 50% washed air losses	46 EER
Indirect cooler with refrigerative second stage	20.2 EER

For comparison, most air cooled refrigerated air conditioning averages about 6-10 EER; cooling-tower versions possibly reaching 15.

Tubular Heat-Exchanger Systems

The Vari-Cool Division of H & C Industries, Santa Rosa, CA, makes heat exchanger cores about 5 ft long, 3 ft wide, and about 2 ft high (1.5 x 0.9 x 0.6 m), containing 165 longitudinal 1 in. (2.4 cm) diameter polystyrene tubes covered with wick-like synthetic fabric sleeves to distribute impinging water and create 100% wetting.

Because the thermal resistivity of dry air films inside the tubes is so great, using plastic instead of metal tubes creates negligible loss, while preventing all corrosion.

Filtered primary air enters the tubes at one end. Secondary air is drawn horizontally around and between the wet-sleeved tubes, in crossflow to the primary air, by propeller fans, and discharged. Small pumps are the only moving parts. Induction primary-air fans attach to the cabinet ends.

The cores are designed to cool 1,000-2,000 cfm (22.7 to 45.4 m^3/min) each of fresh primary air. Up to eight cores can be stacked in cabinets upon standard bases and with standard water distribution pans and covers on top. Two such cabinets side-by-side thus can cool 32,000 cfm (906.1 m^3/min).

These units are usually staged with direct cooling units after the fans. As indirect-direct systems these deliver washed air at about 25% of refrigerating power costs. To gain colder and drier air, three-stage indirect-direct systems with one indirect cooler cooling the

Survey of the Indirect Evaporative Cooling Field

Figure 10-9. A Vari-Cool tubular heat-exchanger. It has five cores stacked over a water-sump base and pump. Primary air enters the tubes, right, and is cooled by descending water exposed to secondary air drawn left-to-right between and around the fabric-covered plastic tubes.

Figure 10-10. The outlet of a Vari-Cool direct second stage. Shown are the pump, left: float valve, right; and thick regid-media cellulose saturating pad, rear.

secondary air for a separate indirect-direct unit correspondingly costs 30%, and three-stage ones with one indirect cooler pre-cooling half of both primary and secondary air for an indirect-direct unit costs 24.8%.

Because washed air sacrifices perhaps half its cooling potential in discharge, these percentages should be doubled for realistic comparisons. However, it is clear that 50% power savings are easily attained, not counting maintenance and possible winter heat savings.

The *Diperi Manufacturing Corp.* of Northridge, CA, also uses fabric-covered polystyrene tubes in horizontal-flow indirect coolers. These have over 200 3/4 in. (1.9 cm) diameter, 3 ft (0.91 m) long tubes in almost cubical cabinets which mount and connect very flexibly in various installations. Each cools up to 4,000 cfm (1.9 m^3/s) of primary air while half as much secondary air is drawn cross-flow between the tubes by a propeller fan.

Performance varies with weather, but reportedly averages 2 to 3 tons (24-26,000 Btuh or 25.3-38 kJ/h) of sensible cooling from less than 1 hp fan and pumping costs. Primary air is cooled about 40% of wet-bulb depression; two units in series raise this to 60%.

Figure 10-11. A basic Diperi indirect cooling installation. Primary air enters the tubes as secondary air enters between the wet fabric-covered plastic tubes, center, propelled by the primary-air fan, far left, and second-air fan, far right. Such units reportedly cool up to 4,000 cfm (1.9 m^3/s) of primary air, 3 tons of sensible cooling for less than 1 hp.

Figure 10-12. Four Diperi indirect coolers in series-parallel. Two pairs in series are mounted side-by-side with a joint secondary-air exhaust plenum between them, exhausted by four rain-hooded propeller fans blowing upward. Primary air enters far right; is cooled, right and center; enters a rotary wheel direct cooler, left center, and is delivered by a fan, left.

Energy Labs, Ltd., of Santa Fe Springs, CA, makes compact single-stage, vertical-tube indirect coolers with enclosed primary air induction blowers. They have banks of closely spaced plastic tubes, about 1 in. (2.5 cm) square, in which sprayed water films slide downward, counterflow to rising secondary air drawn by propeller fans above. Simultaneously, filtered primary air enters between the tubes, which are arranged for maximum turbulent air contact, and the built-in fans deliver it.

The basic model cools either 4,000 or 5,000 cfm (1.9 or 2.4 m^3/s) of primary air, depending on fan sizes; with two tube banks in parallel, outputs are doubled. Primary air cooled 92% by two banks in series often gives excellent comfort single-stage.

These coolers often pre-cool make-up air for large refrigerated cooling systems. Computer simulations of 500-ton (6,330 kJ/hr) combined equipment in 12 widely scattered U.S. cities, including Boston, New York and Chicago, showed average savings that paid for the indirect equipment in only 2.1 years. Naturally, these coolers can be staged with direct coolers as well.

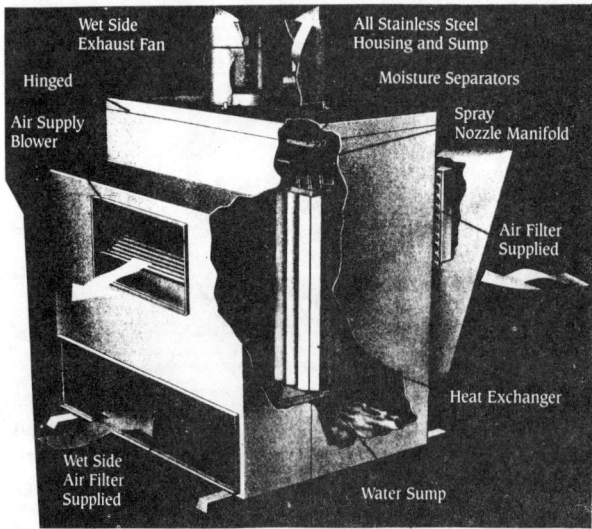

Figure 10-13. A compact Energy Labs vertical-tube indirect cooler. With built-in induction-type primary-air fan, it cools primary air 60% of wet-bulb depression; with two tube-banks in series, 92%. With two in parallel, output is 10,000 cfm (4.8 m³/s).

Rotary Heat-Exchanger Coolers

These were invented in about 1950 by Neal Pennington of Tucson, who used a thick porous wheel of wire screening and aluminum shavings. Cold washed air was blown directly through one half as it revolved slowly. The cooled metal fill revolved into the path of warm primary air which it cooled before revolving back into the washed air.

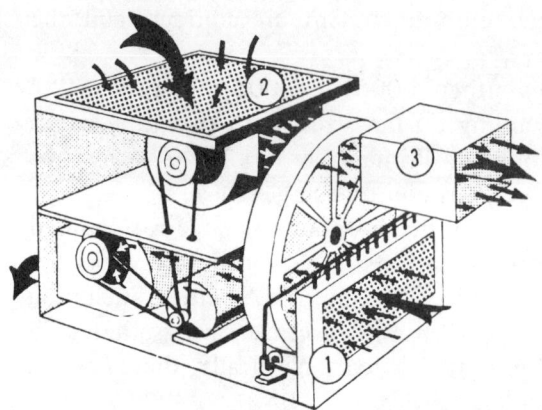

Figure 10-14. An indirect cooler with rotary heat transfer wheel. Washed air from wet pad 1, cools half the turning wheel's porous metal fill. This revolves up into the path of primary air from filter 2 and cools it. Duct 3 delivers it.

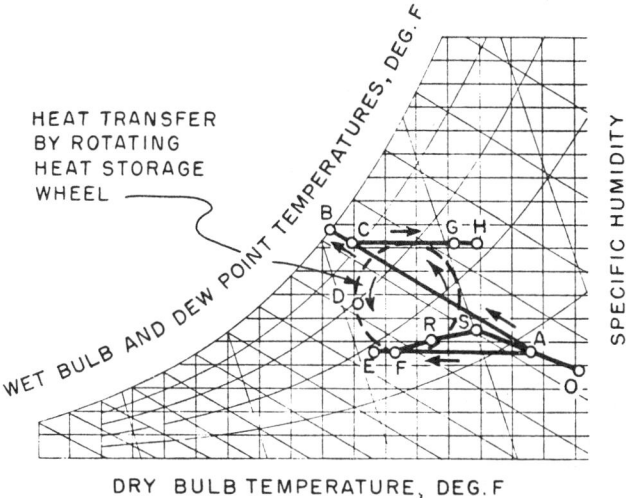

Psychrometric Process in Pennington Indirect Evaporative Coolers. Outside air O and return air S mix, forming air A which is cooled adiabatically to C. It cools wheel filler to D and is warmed to G and exhausted to H. Filler cools dry air from A to E, which enters room at F and absorbs heat to S past room average R. Heat loss AE equals heat gain CG. (Not to scale.)

Figure 10-15.

Today, Norsaire Systems, also maker of cooling-tower and plate-type systems, makes similar large-size systems. Wheels are now spirally wound of alternate flat and corrugated metal ribbons sometimes 10 in. (2.5 cm) wide. Wheels can be 14 ft (4.3 m) diameter if needed.

Highly efficient direct cooler pads blow washed air through half the face while primary air is drawn counterflow through the other; the coldest metal fill thus meets the coolest primary air, maximizing performance.

The cooling efficiencies range between 60 and 90%, averaging about 76%, very good for indirect coolers. However, the wheels have considerable air flow resistance, requiring either extra fan power or the use of very large wheels or several in parallel. Further, considerable air leakage occurs at rim edges and at the dividers between the two air streams.

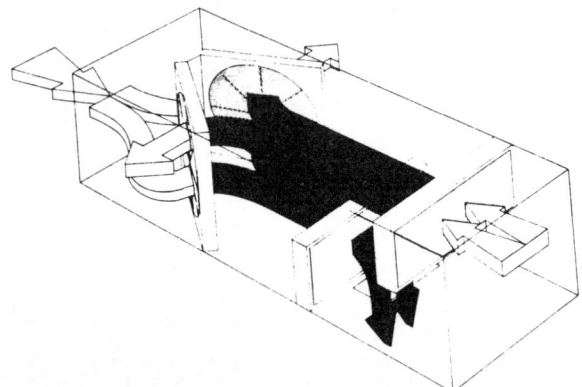

Figure 10-16. A large Norsaire indirect-direct cooler with rotary heat exchanger wheels. Secondary air enters right front, is cooled in a wet pad, the exhausts through and cools two rotary heat exchangers, left center. Primary air enters left rear, is cooled dry by these wheels, then passes through a direct cooling pad, and delivers downward for cooling service, near right.

This manufacturer makes both single stage and indirect-direct models, mostly for 25- to 50-ton (317 to 633 kJ/m) loads. Because of ready wheel availability, some engineers and contractors may build systems. Rotary heat exchangers also serve *desiccant type* coolers noted later.

Plate-Type Indirect Coolers

These typically cool primary air in thin flat tubes called "plates," often of sheet aluminum, exposed on either side to moving secondary air, water films and spray. They were developed by Californians in the 1930s.

The first were failures: huge air-to-air plate-type heat exchangers installed to cool a Walt Disney studio and a medical building received dry washed secondary air from commercial air washers. Stagnant air films insulated both sides of each plate wall and minimized heat transfer, so the equipment was scrapped.

The lesson was clear: primary air must be turbulent enough to prevent stagnant films, and secondary air must contact water films or impinging spray on heat transfer surfaces.

Survey of the Indirect Evaporative Cooling Field

Later designs used tall vertical hollow plates, in which primary air moved counter-flow to descending secondary air and water outside them. Many units opened above so that scale could be brushed from the plates without interrupting operation.

One which cooled the Tucson J.C. Penney store from 1935 to 1955 had 20,000 sq ft (1,858 m²) of plate surface and delivered about 60 tons (760 kJ/hr) of sensible cooling from about 19½ hp. This is equivalent to a 49.5 EER today. When the author examined it in 1952, it operated well, also discharging its secondary air to cool the Penney attic. (See Figure 10-1.)

In 1952-54 for the Navy, the author built and tested three similar plate-type units, including a two-stage one, at the University of Texas at Austin. With the late Professor R.A. Bacon, he derived early design equations published in the author's 1963 book, *Evaporative Air Conditioning*.

A fourth cooler was built by a local engineer for the author's completion. Wetting between its deep, long, closely spaced plates proved difficult, but its 402 sq ft (37.3 m³) of smooth metal cooled primary air 72% of wet-bulb depression, with an equivalent 17.9 EER single-stage economy. It remains an inexpensive design for cooling make-up air. Modern dimpled and absorbent-coated plates might treble its performance.

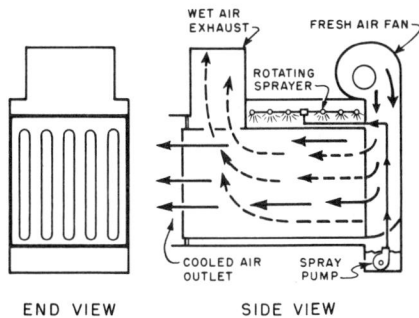

Figure 10-17. An economical design for cooling make-up air. Tested by the author in 1953, it achieved an equivalent of 17.9 EER. Dimpled and absorbent-covered plates might treble performance today.

Today, Des Champs Laboratories, Inc., East Hanover, NJ, makes a wide range of indirect coolers with hollow plates of dimpled aluminum, outputs ranging from 1,700 to 17,000 cfm (48-481 m^3/min). Cooling effectiveness approximates 80% of wet-bulb depression, 81% in larger units.

A typical model with a direct second stage achieved an EER of 76. When reduced 50% for anticipated washed-air losses on discharge, this becomes 38 EER, almost 5 times that of most refrigerated cooling. Power savings of 79% and payback within 3 years were anticipated, not counting reduced compressor wear and possible winter fuel savings.

The larger units are available as complete rooftop modular packages containing, as needed, special filters, heating coils, cooling coils, etc. Various stagings are possible.

Norsaire, noted earlier, also makes large plate-type coolers. In the 5- to 20-ton (63 to 253 kJ/hr) range, these have thin, deep vertical plates with corrugated aluminum surfaces coated with a wick-like absorbent that ensures 100% wetting. Primary air moves horizontally within, cross-flow to rising secondary air and falling drizzles of water.

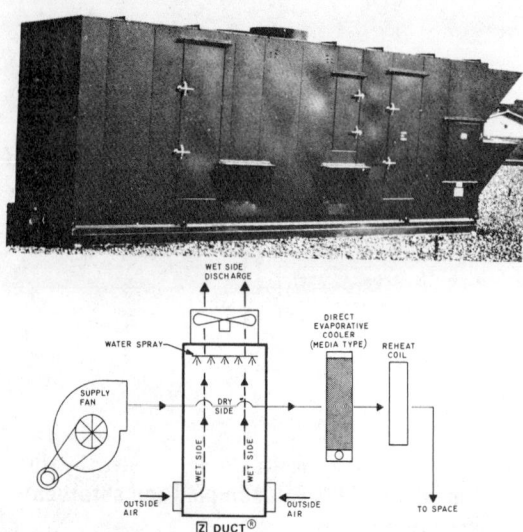

Figure 10-18. A big Des Champs Laboratories indirect-direct cooler with flow diagram. Primary air enters, left, is dry-cooled, center, direct-cooled in a thick wet pad, passes through a winter-use reheat coil, and is delivered, right. Secondary air enters below and discharges above.

Survey of the Indirect Evaporative Cooling Field

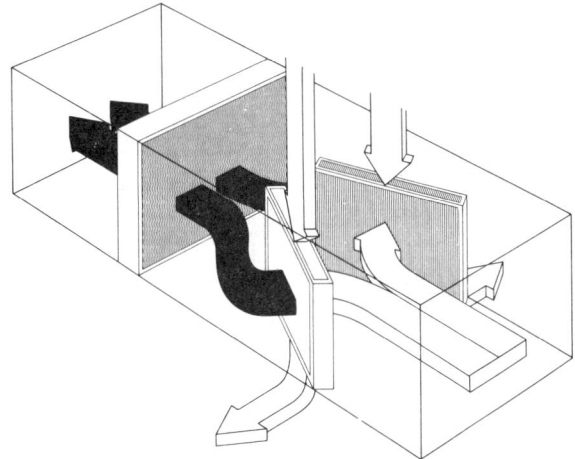

Figure 10-19. A large Norsaire plate-type indirect-direct cooler. Primary air enters, right; is cooled in two plate-type heat exchangers, center right; is cooled further in a direct saturating pad, center left, and is delivered, far left. Secondary air enters above and discharges below.

Reported power savings are 1.4 kW/ton for buildings using 15% make-up air, and 2.2 kW/ton for those using 100% new air.

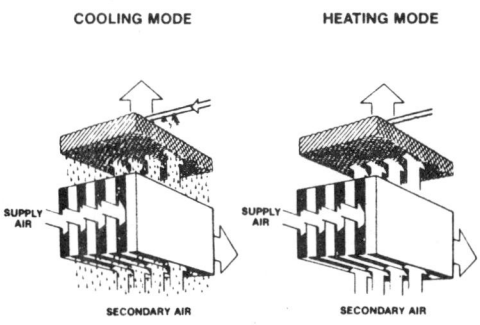

Figure 10-20. Schematic views of Norsaire plate use in summer and winter. This illustrates how easily most indirect coolers transfer waste heat from discharged ventilating air to incoming fresh make-up air.

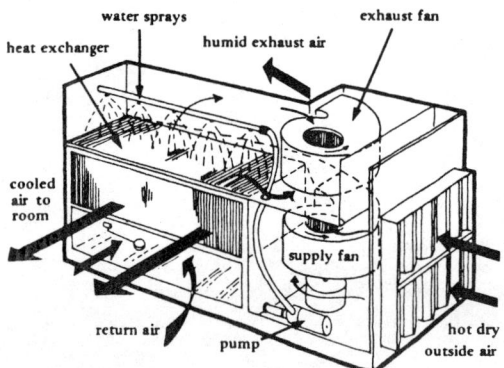

Figure 10-21. An Australian plate-type single-stage cooler. Outside air enters, right, and divides; primary air passes through the heat exchanger horizontally right-to-left, and is delivered indoors. Mixed return and outdoor secondary air are drawn up the heat exchanger vertical passages, counterflow to water films, and discharge, top center.

Commonwealth Scientific and Industrial Research Organization, Melbourne, Australia, developed highly engineered indirect coolers with thin dimpled polyvinyl chloride plastic plates only 2 mm (0.08 in.) apart. The dimples create high air flow turbulence on either side, resulting in primary air cooled 85% of wet-bulb depression.

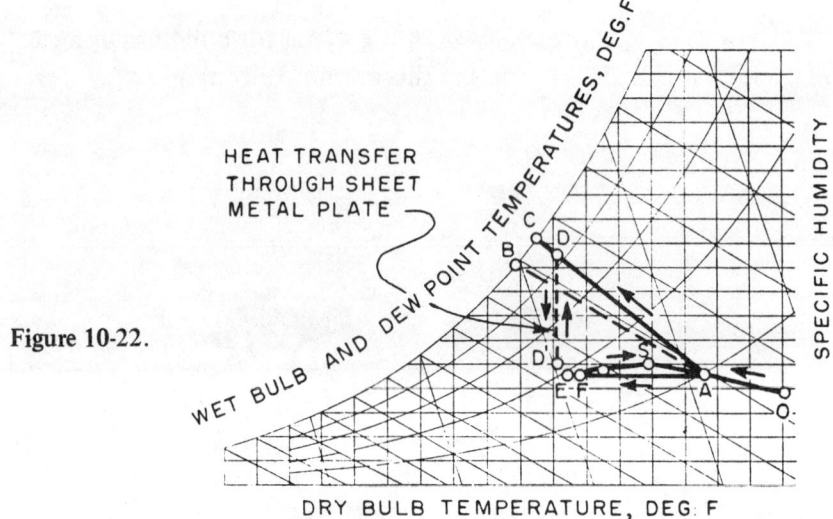

Figure 10-22.

Psychrometric Process in Plate-Type Indirect Evaporative Coolers. Outside air O and return air S mix, forming air A of wet-bulb temperature B. Air A cools sprayed plates nonadiabatically to DD[1], cooling dry air inside from A to E. This enters rooms at F and absorbs heat to S past unmarked room average. Heat loss AE equals heat gain AD. (Not to scale.)

Survey of the Indirect Evaporative Cooling Field 211

Figure 10-23. Australian single-stage plate-type indirect coolers mounted outside telephone switch-gear building. Supply-air ducts enter through windows; return air, through grilles below them.

These coolers are manufactured single-stage to cool hot Australian telephone switch-gear rooms, where washed air might be damaging and refrigeration uneconomic. All return air from the cooled spaces becomes secondary air, so regenerative gains are great. In other usage, staging is possible.

P.H.E. Pty., Ltd., Marleston, South Australia, makes two sizes: 0.5 and 1.5 m^3/s (1,059 and 3,180 cfm). Others may be developed. A unit in Phoenix, AZ, achieves 4.8 tons and an EER of 31.8. Use for winter heat recovery seems rare.

Arvin Air Division of Arvin Industries, Phoenix, AZ, a large drip cooler manufacturer, also makes moderate-size direct coolers whose single thick rigid-media saturation pads offer much better performance and life. Arvin has developed a compact indirect cooler to pair with these direct units in an indirect-direct package.

The heat exchangers are of thin polystyrene sheets embossed and cemented together to form horizontal primary air passages and about 1/8 in. (3.2 mm) thick vertical tubes for secondary air and water.

These tubes are lined with absorbent synthetic "flocking" to hold water films. Water is fed each tube through special wicks from shallow supply troughs above, thus minimizing pumping costs. The films descend in the tubes, meeting secondary air drawn up by propeller fans above. Primary air is drawn through its passages by a centrifugal fan in the direct cooler cabinet.

The two-stage packages are available in four sizes from 2,270 to 3,320 cfm (1.1-1.6 m^3/s), the first stages cooling primary air about 43%. When colder and drier washed air is desired, as in semi-humid areas, two indirect units can feed one direct unit, to achieve 3,900 cfm (1.8 m^3/s) output and dry-cooling of 53%. Either way, washed air is delivered at or below original wet-bulb.

Figure 10-24. An Arvin two-stage indirect-direct plate-type cooler. Air enters the front louvers and divides; primary air successively advances through the heat exchanger, the direct stage saturation pad and its centrifugal fan, rear, and delivers horizontally or downward. The secondary air is drawn up through the heat exchanger, and is discharged by the propeller fan above.

A computer projection of these standard two-stage units averaged over nine southwestern cities' summer weather showed outputs equivalent to 5.4 tons, and EER's of 22, adjusted for washed air loss. Anticipated yearly power savings over 8.9 EER refrigeration averaged $372; over 6.9 EER refrigeration, $496. These savings do not include others in compressor wear and overall maintenance. Use for warming winter make-up air is unreported.

Survey of the Indirect Evaporative Cooling Field

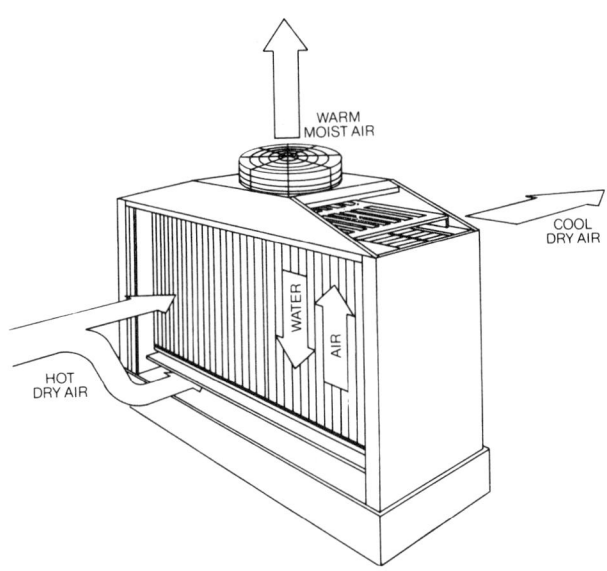

Figure 10-25. Inside an Arvin plate-type indirect stage. Primary air enters the heat-exchanger, left-to-right, as secondary air enters below and is exhausted by the fan above. Water films in the secondary air tubes are supplied by wicks from small supply troughs. The films descend counterflow to rising secondary air.

6500 MasterCool 2-Stage Performance And Savings

CITY	1% ASHRAE DESIGN CONDITIONS		PERFORMANCE AT DESIGN CONDITIONS* DISCHARGE TEMP				ANNUAL COOLING COST SAVINGS***RELATIVE TO A/C WITH EFFICIENCY OF:	
	DRY BULB	WET BULB	MASTER 2-STAGE	CONVENTIONAL COOLER	EQUIVALENT A/C TONS**	MASTERCOOL 2-STAGE S EER	8.9 EER	6.9 EER
PHOENIX	109	71	69.8	78.7	3.1	33	$331	$520
TUCSON	104	66	64.2	73.8	5.2	40	533	687
DENVER	93	59	56.8	66.1	7.9	54	240	327
LAS VEGAS	108	66	64.0	74.6	5.2	47	603	758
SALT LAKE CITY	97	62	60.1	69.3	6.7	51	329	437
ALBUQUERQUE	96	61	58.9	68.2	7.1	50	338	447
EL PASO	100	64	62.2	71.4	5.9	44	507	661
LUBBOCK	98	69	68.1	75.0	3.8	33	233	315
FRESNO SAN BERNARDINO	102	70	69.0	76.6	3.4	39	237	311

*0.3 INCHES EXTERNAL STATIC
**SENSIBLE HEAT RATIO 0.83
***KWH COST OF $.07, THERMOSTAT SET AT 80°

Figure 10-26. Arvin indirect-direct cooler performance in southwest cities. Output averages 5.4 tons of sensible cooling, with washed air 9 deg F (5 deg C) below direct coolers. Average EER, adjusted for washed-air losses, approximates 22. Yearly savings over *good* and *ordinary* refrigeration average 60% and 69% respectively.

The William Lamb Co., of North Hollywood, CA, sells these Arvin indirect-direct packages with photo-voltaic solar panels and accessories to operate them with solar electricity.

A Desiccant-Type Indirect Cooler

The American Solar King Co., of Waco, TX, markets a clever solar heating system, which provides all-year domestic hot water; winter warm-air heating; and summer cooling which uses both a rotary heat-transfer wheel and a desiccant wheel, which resembles the former but is filled with a porous hygroscopic chemical.

Cooling occurs thusly: primary outside air is drawn in and blown through half of the desiccant wheel and gives up humidity. This heats the air, so it is cooled in half of the heat transfer wheel. Now cool and ultra-dry, it passes through a direct cooling pad, emerging as extra-cold washed air for cooling purposes.

After cooling its assigned rooms, it returns as cool secondary air. It passes through the heat exchange wheel, cooling it and warming itself. Then it passes through a solar heated coil connected to solar rooftop collectors. This heats it enough to regenerate (dry) the desiccant. It passes also through the latter, absorbing its active moisture, and discharges outdoors.

This ingenious sytem reportedly provides enough 58 F (14.4 C) washed air to cool 1,800 sq ft (167 m^2) of building, 75,000 Btuh (1,319 kJ/min) of winter heat, plus 90% of hot water needs.

Without energy storage for sunless periods, backup heaters are suggested. However, all-year savings should be great. Federal income tax credits for solar-related installations expired in 1985, but some state ones continue, discounting first costs.

COMFORT AND CLIMATIC ASPECTS

Although human comfort involves more factors, the ASHRAE official Comfort Zone is a standard. However, even when extended for 350 fpm (1.8 m/s) room air velocities, it is less appropriate for evaporative cooling than an older one the author revives in *Evaporative Air Conditioning Handbook*, 1986.

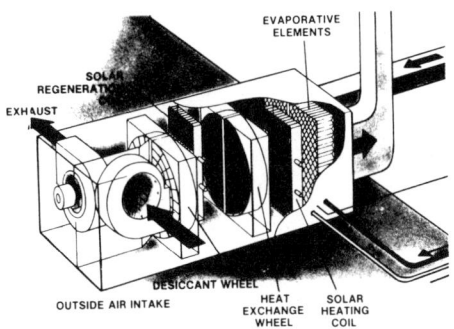

Figure 10-27. American Solar King unit supplies heating, hot water, and indirect-direct cooling. In cooling, a solar-regenerated desiccant wheel, like a rotary heat-exchanger wheel but carrying moisture-absorbing chemicals, dries the air; a rotary heat exchanger dry-cools it, and a saturation pad makes it extra-cold washed air. Heat from roof-top solar collectors keeps the desiccant wheel absorbent.

Nevertheless, most indirect coolers of 60% cooling efficiency with direct second stages can attain ASHRAE comfort when outside wet bulbs are 72 F (22.2 C) or below, provided that related dry bulbs are under 115 F (46.1 C).

The revived evaporative cooling Comfort Zone reaches higher on psychometric charts. Most indirect-direct coolers can correspondingly achieve it when wet bulbs are 78 F (25.6 C), a great geographic advantage. Thus, such coolers can achieve comfort almost everywhere; if not, excellent relief cooling.

Of course, refrigerative second stages open all climates to indirect coolers.

Direct coolers would require wet bulbs below 67 F (19.4 C) and 72 F (22.2 C), respectively, to achieve these zones, limiting regional expansion.

CONCLUSIONS

A wide variety of successful indirect coolers is available, most with second stages. Almost all will operate successfully on, say, 95% of U.S. soil. Almost all save 50% or more of refrigerative system power, plus possible winter fuel.

11

Theory Vs. Practice In Evaporative Roof Spray Cooling

J.C. Smith

INTRODUCTION

Evaporative roof spray cooling systems are designed for installations on the roofs of industrial and commercial buildings. By means of a programmable controller, the systems automatically and uniformly distribute very small amounts of water onto the building's roof. [*Note:* In some situations, the evaporative roof spray cooling system may also be used to mist a building's wall(s).] This is done in such a fashion as to optimize the heat dissipation qualities of the evaporating water film, and thereby to reduce air conditioning usage by between 20-25% with minimal water usage. In buildings without mechanical air conditioning, evaporative roof spray cooling systems will reduce the interior temperature of the facility by as much as 10 deg. MRT. Evaporative roof spray cooling systems also have a positive impact on roof life and maintenance, as attested to by several roofing products manufacturers.

DESIGN, INSTALLATION & COST

The roof surface to which the water is applied may be made of any type of material, and may either be flat or sloped. The water is distributed onto the roof by means of a series of copper pipes and low volume sprayheads.

The water used by the system may come from almost any source; typically, city, well, or waste water is used. The water pressure required (50 p.s.i.) is low enough to preclude the need for pressure-boosting pumps or holding tanks. The temperature of the water is of little importance, as it is the latent heat rather than the sensible heat that determines the cooling effect of the system.

The amount of water used by the roof spray cooling system is controlled by sprayheads through varying their orifice sizes and spray angles, by various forms of temperature and climate sensors, and by the controller's sequence control panel. The sprayheads cover 100 sq. ft. of surface, making it possible to install a wide piping grid utilizing a small amount of pipe, thereby allowing the roof to be easily traversed.

During the installation of the piping system, the roof membrane should not be penetrated. In most situations, the system is attached to the roof by means of blocks and pipe hangers which are affixed to the roof by inorganic adhesive. This permits the entire system to be quickly and easily removed from the roof, if necessary.

The cost of a typical roof spray cooling system varies with the size and/or complexity of the roof; however, the cost to the customer generally averages $.30/sq. ft., factory installed.

The reduction in cooling load will vary from building to building, but generally is significant enough for buildings with roof to floor ratios of 1:1 to 1:2 and with roof insulation of less than R-10 to warrant consideration.

DETERMINING SAVINGS

The limits of the effect of a roof spray cooling system can be determined by isolating the roof (or wall) in question in a cooling load calculation. The most accurate method employed of determining potential savings can be found in the cooling load temperature differential calculations as set forth in the 1981 ASHRAE (American Society of Heating, Refrigeration, and Air Conditioning Engineers) Fundamentals Handbook. While this is a rather laborious process, it yields excellent projections of benefits and costs. Simplified methods of estimating cooling load reductions are practiced, but these ignore the changing solar load over the course of the day and over the course of the cooling season.

In addition, the cooling load temperature differential calculations take into consideration a number of factors, each of which has an impact on the effectiveness of the roof spray cooling system.

These factors are:

1. The intensity of the solar radiation impinging on the roof's surface. This, in turn, depends on 3 factors:
   a. The Time of Day;
   b. The Latitude; and
   c. The Time of Year
2. The construction of the roof and its:
   a. U factor;
   b. Color, and
   c. Rural or urban location
3. The interior of the building — its:
   a. Design temperature; and
   b. Plenum or false ceiling
4. The efficiency of the present mechanical air conditioning equipment and its operating hours
5. Electrical power consumption and its cost

And, finally,

6. Water — its supply, quality, and cost.

THERMODYNAMIC PRINCIPLES

The ASHRAE-based roof spray cooling calculations and the phenomena they describe are based on two basic thermodynamic concepts: (1) The Latent Heat of Vaporization of Water; and (2) Heat Flow.

The Latent Heat of Vaporization

One attraction of roof spray cooling lies in its conceptual simplicity. Using water as a coolant is certainly neither mysterious nor unique; all of us perspire, and physiologically derive benefit from this effect. Moreover, few of us have not experienced the chilling effect of direct evaporative cooling upon stepping out of a swimming pool, or even out of a shower. It is the ability of water to absorb a large amount of heat during vaporization that creates this cooling effect, and, in fact, is the basis for the effectiveness of roof spray cooling.

Water exists in 3 phases: as a solid, as a liquid, and as a gas. At one extreme, as water changes from a solid to a liquid (the Latent Heat of Fusion), the amount of energy absorbed is only *80 calories per gram*. At the other extreme, when the liquid is vaporized, the amount of heat absorbed in the transformation (the Latent Heat of Vaporization) is *537.7 calories/gram, or 7 times* the heat absorbed during fusion.

Translating this into more familiar terms, at 212°F, the heat absorbed by 1 pound of water is *970 BTU's, or 8,080 BTU's per gallon.* At lower temperatures—for example, 120°F—the heat absorption characteristics of water are amplified, and the heat exchange is increased to *1,025 BTU's/pound,* or approximately *8,538 BTU'S per gallon* of water evaporated.

Rather than confining this physical phenomenon to the inside of coils and tubes, as with traditional mechanical air conditioning, roof spray cooling applies it directly to the largest source of external heat on the typical one- to two-story commercial/industrial building by spraying the roof with a fine mist of water and allowing the water to evaporate completely.

The result, with proper design and application, is that a roof no longer acts as one huge solar heat absorption panel. Instead, roof spray cooling allows a roof to remain at or *below* the ambient temperature. This means that the heat load contributed by the roof is negated; moreover, at certain times of the day, the roof will actually absorb heat from the interior of a structure and release it to the outside.

In order to appreciate the magnitude of the phenomenon of the Latent Heat of Vaporization as it applies to roof spray cooling, let us consider the following example:

1. The roof surface temperature of our building is 120°F;
2. A gallon of water weighs 8.33 pounds;
3. The roof surface area is 10,000 sq. ft.;
4. The roof spray cooling system is operational for 8 hours a day; and

5. The solar load on this day is constant (assumed for convenience).

This roof can evaporate approximately 1,000 gallons of water over the course of this day, or 8,330 pounds of water. Assuming a roof temperature or 120°F and a conversion value of 1,025 BTU's/pound, this means that over *8.5 million BTU's* (8.330 lbs. X 1,025 BTU's/lb.) will be removed by this evaporation on the surface of the roof.

Note, of course, that these numbers do not imply that all of this heat would have penetrated into the building if unsprayed. Much of the heat would have been absorbed by the roof and roof insulation and/or reflected. However, it should also be noted that this heat absorption does have a deleterious effect on the roof structure itself.

HEAT FLOW

Heat always flows from a warmer to a cooler environment. Consequently, heat will flow into a building during the summer, while during the winter, heat will flow from the building to the outside. Such heat flow will cause the temperature of the interior air to rise as long as it is not removed (through the use of mechanical air conditioning) as fast as it flows in.

Variables Affecting Heat Flow/Gain: The total rate of heat flow or heat gain depends on three factors:

1. The heat conducting properties of the material through which the heat is passing:
2. The total area of the material through which the heat is flowing; and
3. The difference in temperature between the warmer and cooler sides of the material.

These factors are expressed in the standard heat flow calculation:

$$q = AU(t_2 - t_1)$$

Where q = heat gain in BTU's/hour

A = area

U = conductance

$t_2 - t_1$ = the difference in temperature between the outside & the inside surface

Conductance: We typically refer to the U value or R factor when speaking of the insulating properties of a material.

The U value signifies the *conductance* of heat through a non-homogenous material (such as a roof) in terms of BTU's/hour through 1 square foot of that material (of a specified thickness) for a one degree temperature differential.

The R factor, or resistance factor, is simply the inverse of the U value. For example, if the U value = .20, the R factor is 5; a roof evaluated at R-10 means a U value of .10. Most roofs of industrial buildings have a U value of between .30 and .10, by design. A bare metal roof will have an R factor of 0, while the same roof with 1-1/16" of fiberglass insulation will have an R factor of approximately 4.25.

How U Values and R Factors Are Determined:

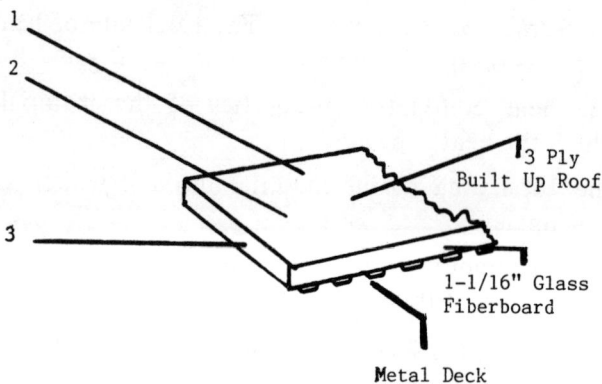

Figure 11-1. Determining U Values and R Factors

Item	Description	Resistance R factor	Conductance U Value
1.	Outside air film (7.5 mph wind velocity)	0.25	4.00
2.	Built Up Roofing	0.33	3.03
3.	Insulation (x/k, 1.0625/.25)	4.25	0.235
4.	Metal Deck	0.00	0.00
	Total: (U = 1/R)	**4.83**	**0.207**

Area: The roof surface of a one- or two-story building is the building's largest source of heat gain. This is easy to understand by considering the following example.

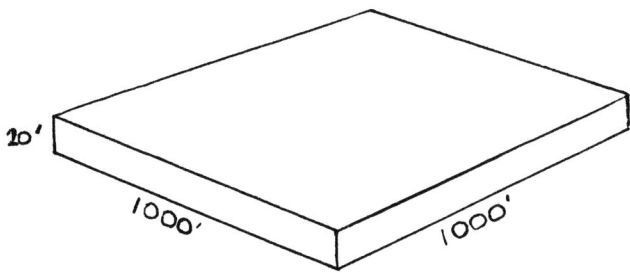

Figure 11-2. Roof area example

As illustrated above, a building with a roof/floor area of 1,000,000 sq. ft. (or 1,000 sq. ft. long by 1,000 sq. ft. wide) has only 80,000 sq. ft. of wall area (1,000 ft. long by 20 ft. high X 4 sides), or 12.5 times as much roof as wall area. Moreover, the roof area is always being affected by solar radiation, while the east- and west-facing walls of the structure are each only directly affected for half a day, and the north- and south-facing sides of a structure are never affected directly by solar radiation. And, obviously, the larger the facility, the greater the amount of heat entering the building.

Temperature Differential: The heat flow through roofs (and walls) depends on the temperature difference across the roof (or wall) from the outside surface to the interior surface.

Surface Temperatures: The radiation impinging on the roof's surface is directly related to the angle of the sun's rays over the course of the day. As the sun's rays hitting the building's roof approach an angle of 90°, the amount of radiant energy is maximized.

By the same token, the latitude and time of year also have a bearing on the angle at which the sun's rays are hitting the roof.

Obviously, the solar radiation strikes more squarely on the roof at the lower than at the higher latitudes. In addition, the amount of radiation will vary as the earth revolves and changes its orientation (and tilt) in relation to the sun on a daily/monthly basis, which results in the various seasons.

The fact that the earth and the building's roof change their orientation to the sun on an hourly basis will complicate our heat flow calculations considerably. However, for our *simple* heat flow calculation, we will assume that the sun's radiation is constant.

What is always considered a constant in both our simple and more realistic calculations is the interior building temperature.

Interior Temperature: The interior design temperature is that temperature at which a facility operates at maximum comfort and efficiency. Generally, the design temperature is set at 78°F, although some operations may require much lower interior temperatures (such as computer rooms, cold storage areas, etc.), while other operations may be able to operate at interior temperatures higher than 78° (warehouses, for example).

Static Heat Flow Calculation — A Simple Example: Let us now consider an example of how much heat will enter a building during the course of just one hour. In our calculations, we'll consider a 50,000 sq. ft. light manufacturing facility, with roof construction as shown under "Conductance" above (i.e., this roof has a U value of .207). If the temperature of the top side of this roof is 140°F, and the temperature is 78°F on the inside, how much heat will flow through the roof in one hour?

The calculation is as follows:

$$q = AU(t_2 - T_1)$$

Where A = Area
U = U Value
t_2 = Outside Temperature
t_1 = Inside Temperature

$$q = 50,000(.207)(140 - 78)$$
$$q = \textit{641,000} \text{ BTU's/Hour}$$

If 12,000 BTU's equal 1 ton of mechanical air conditioning, then it would require *53.5 tons* of air conditioning to displace the amount of heat that enters our example facility in one hour.

Solutions to Heat Gain

The solutions to heat gain are: 1) increase the resistance to the flow through the use of additional insulation in the roof; 2) increase the amount of air conditioning, either through additional tonnage or lengthening the "on" times to pump out the heat faster than it can come it; 3) decrease the temperature differential from the outside of the roof to the inside by adjusting or allowing the internal temperature to increase, say, from 78°F to 85°F; and 4) decrease the temperature differential across the roof from the exterior to approach the ambient wet bulb temperature through the use of a roof spray cooling system.

Roof Insulation and Solar Gain: The purpose and function of insulation is to impede or slow the flow of heat from the outside surface to the interior. A wall (or roof, for that matter) with sufficient thermal mass would actually completely block the heat flow from the outdoors. This, however, is never done, simply because, structurally and economically, it would not make sense.

Using our example above (roof temperature 140°F, interior temperature 78°F), it would take *64 feet* of concrete, or *29 inches* of fiberglass insulation to create sufficient resistance to *completely* block the incoming heat flow.

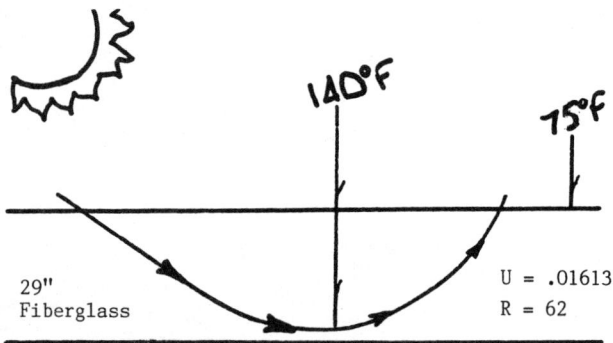

Figure 11-3. Insulation required to block heat gain.

The effect of insulation during the summer months is *not* to prevent all of the heat flow, but rather to delay it from entering the building's interior for a period of time.

Mechanical Air Conditioning: Mechanical removal of heat through the use of air conditioning units is a method that doesn't involve the heat flow calculation itself, but determining the amount of air conditioning tonnage required to do the job adequately does depend on calculating heat flow. Mechanical air conditioning involves the removal of heat gain at a rate equal to the rate that the heat is entering the building. In our example, the amount of air conditioning tonnage required to displace the heat entering the building *only* through the roof for that one-hour time period would be *53.5 tons.*

Roof Spray Cooling as a Solution to Heat Gain: The final solution would be to reduce the temperature of the surface of the roof to the extent that the temperature differential across the roof is reduced, and therefore the heat gain is reduced. The effect of lowering the temperature differential from 62 degrees (140° - 78°) to 17 degrees (95° - 78°) is to virtually eliminate the heat gain. When viewed over the course of a 24-hour day, the roof spray cooling system, in addition to eliminating the heat gain, can actually draw heat from the interior of the building.

Figure 11-4. Mechanical air conditioning tonnage required to remove heat gain from roof.

DEVELOPMENT OF ROOF SPRAY COOLING

Cooling the interior of a roof through the evaporation of water has been around for some time. The Brazilians used a network of open conduit on their roofs through which water flowed, acting both as a simple air conditioner and as a water heater. The Indians cooled the Orient Express through the use of burlap placed on the train's roof, which was wetted from station to station.

In the U.S., the method has been employed, or rather, recognized, since the 1930's. Dr. Willis Carrier, the acknowledged father of mechanical air conditioning, was a proponent of the method. When asked why he did not use the idea himself, he replied that his business was the manufacture of compressors and related equipment only; even so, Dr. Carrier encouraged the development of roof spray cooling.

Roof spray cooling's relative neglect over the preceding 50 years can be attributed to what might be called a lack of elegance; and designers and manufacturers of roof spray cooling systems have only contributed to the problem by approaching it in a very crude and typically unsophisticated manner.

Initially, the method of distribution employed was impact spray heads or lawn sprinkler heads, which, while more effective than nothing at all, resulted in roof water coverage that was both uneven and excessive. In addition, due to the amount of pressure required to charge such a system with water, pumps and storage tanks were required. The off/on control of these systems was either manual or a simple solenoid/thermostat mechanism, which typically sprayed either too much or too little water.

These findings led to the development of a "punched pipe" system. With punched pipe systems, holes are punched or milled into "sticks" of copper or PVC pipe. These holes typically are 1"-1½" separated and located at the 10 o'clock and 2 o'clock positions on the pipe. When properly charged, these pipes spray water in much smaller droplets than the impact sprayheads, and consequently have a better coverage pattern. Also, because less water per line is used, the use of pumps and storage tanks is usually unnecessary. Until about 2 years ago, the control systems on these punched pipe systems differed little from those used in the '30's and '40's. Some are still sold with a mechanical cam timer system, though generally today an electronic control system is employed.

There are two major drawbacks to these punched pipe systems, discounting the controls completely:

- First, the amount of water sprayed from the pipe is still excessive; studies have shown that approximately twice as much water is used as is necessary. Thus, usually the pressure drop across a line or entire field (unless it is short or relatively small) causes "leakers." This occurs where the water pressure isn't high enough to force the water out of the hole with enough force to cause the water to break into a mist. This results in uneven spray patterns and less than adequate coverage.

- Second, maintenance on these systems is rather involved. The influx of any foreign matter, including such things as sand, clogs holes. In addition, because it is impossible to filter the water just prior to its issue from the hole, and because the hole is punched directly into the copper pipe which oxidizes and captures any precipitates from the water, the holes frequently need to be cleaned with a brush from the outside and annually through swabbing the inside of each pipe.

The resurgence of interest in and use of roof spray cooling systems is a direct result of their evolution over the past few years from inherent theoretical acceptance to effective practical application. Through advances in system engineering and design—chief amoung them being the advent and utilization of "microprocessor" controls and low volume, low maintenance sprayheads—the "low tech" idea of roof spray cooling has been adapted to fit the needs of a "high tech" world. Through sophisticated design, energy analysis, and system control, roof spray cooling has become an eminently viable cooling and energy conservation alternative for the 1980's and beyond.

MAKING ROOF SPRAY COOLING EFFECTIVE

Two specific advances in roof spray cooling technology have served to optimize the effectiveness of roof spray cooling systems, while minimizing the cost of operation.

Sprayheads Designed for Roof Spray Cooling Alone

The design and manufacture of low volume sprayheads developed specifically for roof spray cooling systems solve a number of problems that plague systems utilizing older (circa 1930) punched pipe technology.

These new sprayheads are designed so that the volume of water and the direction of the water mist can be altered in a variety of ways. Each sprayhead can be modified so that it supplies the right amount of water in the correct pattern for its specific location on the system. Earlier roof cooling systems employing punched pipe permitted only one pattern and amount of water flow. Moreover, alteration of the spray pattern or water flow stoppage was often a problem caused by clogging or oxidation of the spray holes; the integrity of the hole could be damaged even during shipping. In addition, due to the fact that these punched holes were relatively large in diameter and were only 16"-18" apart, the fall in pressure across a line and the amount of water used were great. Very seldom did a line spray uniformly across its length.

With the new roof spray cooling sprayhead, water can be directed away from any equipment on the roof top, and can be applied in varying amounts to the roof surface. The individual filters in each sprayhead inhibit clogging of the sprayhead orifice. Unlike the older systems, which often required regular and laborious brushing out of each spray hole, the sprayhead maintenance, if necessary, is as easy as unscrewing the head from its base and back-flushing the filter. Finally, the construction of the sprayheads makes them sturdy enough to be guaranteed for the life of the roof spray system.

Microprocessor-Based Control System

Until recently, the typical roof spray cooling system not only was an arrangement of punched pipe, but also was controlled by a mechanical control that consisted of a number of cams that, once set, were exceedingly difficult to adjust; in fact, they had only one setting, meaning the water schedule stayed the same not only throughout the day, but throughout the entire summer cooling season.

This situation caused considerable overspray and ponding in the relatively cooler months of May and September, or underspray and thus limited effectiveness in the hot months of June through August.

The use of advanced control systems has eliminated these problems by monitoring temperature variation throughout every day of the cooling season and altering the amount of water as the conditions for optimal evaporation change. Thus, only that specific amount of water that can be evaporated at a particular point in time will be sprayed by the system as it runs from day to day throughout the summer. This eliminates ponding and optimizes the cooling effect of the water.

Roof Spray Cooling – A Viable Alternative

What these significant advances mean is the integration of the roof spray cooling system into a facility's complete HVAC system, rather than the system acting as a mere add-on, operating in isolation.

Today's roof spray cooling system so designed becomes a viable alternative to increased roof insulation and/or additional air conditioning tonnage, both in existing and in new construction, as a means of energy consumption control in air conditioned buildings, and as an effective means of increasing worker comfort and productivity, and the integrity of goods and machinery in un-air conditioned structures.

ANALYSIS OF
ROOF SPRAY COOLING SYSTEM TEST RESULTS

Houghton, et al — ASHVE — Pittsburgh, PA

Houghton, Olson, and Gutberlet presented a paper at the semi-annual meeting of the American Society of Heating and Ventilation Engineers in June of 1940. The paper was entitled "Summer Cooling Load as Affected by Heat Gain Through Dry, Sprinkled, and Water Covered Roofs."

The research was conducted by the ASHVE Research Lab in Pittsburgh, PA. A test building was constructed on which the heat flow through various types of roofs with and without several types of treatment were measured. The interior of the building was maintained at a constant 75°F and 50% relative humidity. The tests involving the sprinkled roof were conducted on September 8. The calculations were then corrected for the design day of August 1. The outdoor dry bulb temperature ranged from 77°F to 95°F (the design db Temp.), while the wet bulb temperature ranged from 68°F to 75°F, and therefore the relative humidity varied from 64% to 40% during the day.

The summary of the studies indicated the following as concerned roof spray cooling and the concrete asphalt roof, in terms of heat flow (BTU/SQ. FT./HR.).

TIME	WITHOUT SPRAY	WITH SPRAY
5 a.m.	−2.3	−4.0
6	−2.0	−3.8
7	−1.6	−3.2
8	−0.2	−2.0
9	2.0	−1.0
10	5.5	0
11	9.0	0.7
12 p.m.	12.2	1.1
1	15.3	1.7
2	17.5	2.0
3	18.0	2.1
4	17.0	2.0
5	15.5	1.8

TIME	WITHOUT SPRAY	WITH SPRAY
6 p.m.	13.1	1.0
7	10.1	0
8	7.0	−1.0
9	4.6	−1.8
10	3.0	−2.5
11	1.4	−3.0
12 a.m.	0.3	−3.5
1	−0.5	−3.8
2	−1.1	−4.0
3	−2.0	−4.0
4	−2.3	−4.1

Examination of these figures reveals the following:

Maximum Heat Flow (BTU/SQ. FT./HR):
  Without Spray   18.0
  With Spray   2.1

Minimum Heat Flow (BTU/SQ. FT./HR):
  Without Spray   −2.3
  With Spray   −4.1

Length of Time (in Hours) with No or Negative Heat Flow:
  Without Spray   8 Hours
  With Spray   16 Hours

Average Heat Flow (Per Hour) Over 24-Hour Period:
  Without Spray   +5.81 BTU/Sq. Ft./Hr.
  With Spray   −1.22 BTU/Sq. Ft./Hr.

According to Houghton, et al, "The effect of water in . . . the case of the sprinkled . . . roof is to greatly reduce the rate of heat flow from that found for the same panels in dry conditions. Of greater interest, however, is the effect of the water to absorb a large part of the radiant heat, to retain it . . . and to dissipate it back to the air through the latent heat of evaporation."

Yellott – ASHRAE – Phoenix, Arizona

John Yellott presented a paper to the Association of Heating, Refrigeration and Air Conditioning Engineers in 1966 titled "Roof

Theory Vs. Practice In Evaporative Roof Spray Cooling

Cooling with Intermittent Water Sprays." Rather than build an entire structure and measure the reduction of heat flow through the roof into an air conditioned environment, Yellott constructed an open roof deck and measured the temperature of the deck itself, with and without water spray. These tests were conducted in Phoenix, Arizona, and consequently the ambient conditions surrounding them were much different from those in Houghton's Pittsburgh tests.

Fourteen tests were run between July 17 and August 4, and the August 3rd test was designated by Yellott as typical.

August 3rd Test

[Notes: Col. 1 — Time
Col. 2 — Without Spray (Deg. F)
Col. 3 — With Spray (Deg. F)
Col. 4 — Difference (3 − 2) (Deg. F)
Col. 5 — Air Temperature (Deg. F)
Col. 6 — Difference in Sprayed Roof & Air Temperature (Deg. F)
Col. 7 — Relative Humidity (%)]

Col. 1	Col. 2	Col. 3	Col. 4	Col. 5	Col. 6	Col. 7
5 a.m.	74	74	0	80	−6	50
6	75	75	0	82	−7	
7	98	90	−8	87	3	
8	120	98	−22	90	8	42
9	145	106	−39	92	14	
10	160	110	−50	95	15	
11	172	110	−62	99	11	32
12 p.m.	177	111	−66	100	11	
1	177	110	−67	100	10	
2	170	103	−67	102	−1	20
3	159	98	−61	103	−5	
4	143	90	−53	103	−13	
6	124	82	−42	102	−20	19

These figures reveal the following:

Maximum Temperature of the Roof Surface (°F):
    Without Spray   177
    With Spray   111

Average Hourly Temperature Difference Between Roof Surface & Ambient Air (°F):

    Without Spray   40.92
    With Spray   1.69

As is evident in the earliest Houghton, et al, study, the sprayed roof shifts from a heat panel to a cooling panel much sooner and to a much greater degree than does the unsprayed roof. Unfortunately, Yellott did not continue to monitor this cooling panel effect over a 24-hour period.

Yellott does go on to point out, however, that "evaporating water is by far the most economical refrigerant, since 1 gallon represents about 8,500 Btu of cooling capability and 1,000 gallons, used completely, will produce somewhat more than 700 tons of refrigeration."

Srivastava, et al — Centre of Energy Studies, Indian Institute of Technology, Jodphur, India

Srivastava, Navak, Tiwari and Sodha of the Centre of Energy Studies, Indian Institute of Technology, designed a dormitory for students in Jodphur, India, which incorporated a roof spray cooling system. Their study was published in "Energy & Buildings" (Vol. 6, 1984), and is entitled "Design & Thermal Performance of a Passive Cooled Building for the Semiarid Climate of India."

Several methods of cooling the dormitory were considered, with the most effective being the use of evaporative roof spray cooling in conjunction with "desert cooling fans."

Below is represented the effect of merely the roof spray cooling system on the interior temperature of the uncooled building (that is, operating without the fans):

Time	Ambient Temp. °F	Interior Temp. Room W/O Spray °F	Interior Temp. Room With Spray °F	Room Difference °F
1 a.m.	87.8	97.2	91.6	5.6
3	85.6	96.4	90.5	5.9

Time	Ambient Temp. °F	Interior Temp. Room W/O Spray °F	Interior Temp. Room With Spray °F	Room Difference °F
5 a.m.	81.7	95.4	88.7	6.7
7	85.1	94.1	87.4	6.7
9	91.9	85.0	88.2	6.8
11	96.6	96.3	89.6	6.7
1 p.m.	102.4	97.3	90.1	7.2
3	105.8	98.4	91.4	7.0
5	104.0	99.0	92.8	6.2
7	99.5	98.8	93.0	5.8
9	95.0	98.6	92.8	5.8
11	91.4	97.7	92.3	5.4
1 a.m.	87.8	97.3	91.4	5.9

Electrical Components Manufacturing Plant, Rio Piedras, Puerto Rico

This test was performed by company employees in order to evaluate the effectiveness of the roof spray cooling system recently installed on their facility. It was conducted on two consecutive days in May, 1982 in Rio Piedras, PR. As you will note there was a slight variance in ambient temperature from one day to the next; however, the results still conform closely to the studies previously discussed.

Time	Ambient Temp. Day One °F	Roof Surface Temp. W/O Spray °F	Ambient Temp. Day Two °F	Roof Surface Temp. W/ Spray °F
7 a.m.	78.8	78.1	80.6	80.6
8	82.4	83.8	84.2	81.7
9	86.0	95.0	85.1	80.8
10	86.0	105.3	83.3	82.8
11	86.0	106.3	84.0	83.1
12 p.m.	86.0	107.6	85.0	83.6
1	87.8	106.3	86.0	83.7
2	88.7	106.3	84.2	84.1
3	86.0	105.3	84.0	81.1
4	84.0	100.6	82.0	81.0
5	82.0	95.0	80.0	80.8

A portion of this building was air conditioned and maintained at a temperature ranging from 74.6°F to 80°F during the day, while the tool room and stock room were un-air conditioned, and their temperature varied more widely, depending in the main on the external load, a large part of which naturally was the roof.

The air conditioned portion of the plant was monitored for electrical consumption of the air conditioning units both while the roof spray cooling system was and was not being employed. The un-air conditioned areas were monitored for dry bulb temperature on both days.

Again, because these tests were run in conditions that varied from one day to the next, a *direct* comparison cannot validly be made. However, the recorded data is as follows:

Time	Average Temperature Un-A/C'D Areas Without Spray Day One °F	Average Temperature Un-A/C'D Areas With Spray Day Two °F
7 a.m.	78.8	78.8
8	81.5	82.4
9	84.2	83.8
10	89.6	86.9
11	93.7	87.6
12 p.m.	95.0	85.3
1	95.0	84.5
2	95.0	84.0
3	93.7	83.3
4	85.1	83.3
5	84.1	82.9

Time	Power Consumption (KVA) In A/C'D Area Without Spray Day One	Power Consumption (KVA) In A/C'D Area With Spray Day Two
7 a.m.	128.9	128.9
8	170.4	164.6
9	187.1	173.8
10	188.7	173.8

Time	Power Consumption (KVA) In A/C'D Area Without Spray Day One	Power Consumption (KVA) In A/C'D Area With Spray Day Two
11 a.m.	191.2	174.6
12 p.m.	191.2	174.6
1	199.5	174.6
2	199.5	174.6
3	199.5	158.0
4	191.2	158.0
5	187.1	138.0
Average	184.9	163.0

As noted, while direct comparison of the data gathered from these tests must be undertaken with caution, the reduction in the interior temperature of the un-air conditioned area from Day One (without roof spray) to Day Two (with roof spray) of 10.5°F at 1:00 p.m. can be attributed in large part to the cooling of the roof.

The same can be said of the reduction in power consumption for the air conditioned area, which averaged 21.9 kva/hour difference over the test period.

REFERENCES

Houghton, F.C., Olson, H.T., and Gutberlet, Carl. "Summer Cooling Load as Affected by Heat Gain Through Dry, Sprinkled and Water Covered Roofs." *Transaction American Society of Heating and Ventilating Engineers,* June, 1940, pp. 231-244.

Srivastava, A., Nayak, J.K., Tiwari, G.N. and Sodha, M.S. "Design and Thermal Performance of a Passive Cooled Building for the Semiarid Climate of India." *Energy & Buildings,* Vol. 6, pp. 3-13.

Yellott, John. "Roof Cooling with Intermittent Water Sprays." *ASHRAE Transactions,* 1966, pp. III.1.1-III. 1.8.

SECTION V
HVAC CONTROLS

12

Comparison of Cost and Performance Of HVAC Controls

C.E. Lundstrom

INTRODUCTION

During the past decade, sizable challenges have been faced by design engineers, building owners, and facility operations personnel in choosing the right HVAC controls and Energy Monitoring and Control System (EMCS) for their applications. Recent advancements in HVAC controls and EMCS are going to make those decisions even more complex. Decision makers will need to keep abreast of the ever-changing controls market. The recent introduction of microprocessor-based control systems which are smaller, modular, and less expensive will require engineers, owners, and operators to compare the performance to the cost of HVAC controls when the installation of new controls becomes necessary.

The information provided in this chapter will identify some of the most recent advances in controls, compare the performance of the control options, and compare initial costs and life cycle costs for various alternatives.

RECENT ADVANCES IN HVAC CONTROLS

Direct digital controls (DDC) have been used in HVAC for a number of years and generally would not be considered a recent advancement. However, the size and form of the DDC which are presently appearing in the HVAC market could be termed "state of the art." The current trend in DDC is toward smaller, modular, microprocessor-based controllers. Moreover, manufacturers in increasing number and types are supplying DDC—and especially the HVAC equipment suppliers themselves.

Chiller Controls

Within the last year, manufacturers of large chillers have begun to install sophisticated microprocessor-based controllers on their chillers as standard factory equipment. Essential control and status monitoring points from the entire chiller system are being fed into a single DDC panel. The panels have a data display line which allows the building operator to read out any parameters in English descriptions.

Typical operating parameters which are provided on the controller include:

- Chilled- and condensor-liquid temperatures.
- Evaporator- and condensor-refrigerant pressures.
- Percentage of motor current.
- Oil pressure.

As with other types of DDC, temperature controls on the chiller packages can be either proportional or proportional-plus-derivative. A built-in clock gives automatic start/stop control over both chiller functions, as well as over all the associated auxiliary equipment, including the water pumps and cooling tower.

Built into some systems is the ability to interface directly with building automation systems. The RS-232 connections allow remote control or monitoring of the panels. These types of connections may have limitations, depending upon the ability of the chiller control panel to "interface" with other EMCS/DDC building controllers. The execution of some of the more sophisticated chiller optimization programs (such as those for chiller sequencing, water temperature resetting, and demand limiting) will require a separate processor and the ability to override or change parameters at each chiller. This may mean using a different manufacturer of building automation systems, thus raising the question of interface.

Boiler Controls

Not to be outdone by the chiller industry, the boiler manufacturers have come out with their own version of DDC for hot water boilers. Because the control strategies for boilers are fairly straightforward, their controls do not require the sophistication of a comlex chiller package. Examples of the types of controls developed for boilers include:

- Sequencing of boilers to maintain water temperatures.
- Outdoor, indoor, and water temperature monitoring for hot-water temperature reset.
- Lead-lag capability for increased efficiency and longer boiler life.

A built-in clock provides automatic start/stop control of the boiler system. Like the chiller DDC controllers, boiler controller interface may or may not be possible, depending on who manufactured the control system. This may not be as critical a problem with the boiler system as it is with a chiller system.

HVAC Controls

The types and trends in control systems for and from the HVAC industry is very interesting. First of all, from the control industry, second- and third-generation DDC panels are becoming more powerful in both software and hardware. Many manufacturers are offering lines of DDC panels with fewer control and monitoring points. This is probably in response to the general criticism of DDC reliability, which includes counting on one processor to handle many control loops. One manufacturer has a DDC panel with an on-board diagnostic processor, which they claim can diagnose failures down to the chip level. This means a repair may be as simple as replacing a very inexpensive chip instead of replacing a very expensive motherboard.

Another criticism made of the first generation of DDC systems was that the systems were too hard to program. Trying to compromise on a software programming language which is both "easy to program" and "provides good performance" may be mutually exclusive. Languages designed for ease of programming normally run too slow and can be cumbersome. Software designed to speed up the process requires a more experienced programmer. With the current generation of DDC, manufacturers are trying to come up with their versions of compromise languages. Most of these are based on a common language such as Basic or Pascal, but the controls people have undertaken the necessary modifications to make them operate at reasonable speeds.

While the controls industry was pouring manpower into third-generation DDC systems, the HVAC industry was looking at ways to install microprocessor-based controls as standard equipment from the factory. Many large manufacturers of HVAC systems either have signed agreements with control manufacturers to supply controls or have developed their own control systems and begun shipping equipment with DDC-mounted controls from the factory. The types of functions available for the HVAC controls basically run the same gamut as the pneumatic and electronic controls they will be replacing (coil, damper, and fan controls), but with the added capability to tie them back into a building automation system.

Zone Controls

Probably the newest development in DDC controls within the past year is the increase in the number of companies offering microprocessor-based, zone terminal controls. This concept is not totally new to the industry. But, with the recent introduction of many different and powerful controllers on the market within just the past few months, market analysis is revealing a trend toward this type of design. For many years, the best argument against DDC was the need for having too many control loops maintained on one processor board. If the processor board crashed, many environmental zones could be affected. With the new individual DDC terminal loop controllers, this problem has been alleviated. the overall sophistication designed into the new controllers presents some new design and energy management capabilities not previously available.

Each manufacturer's controller has different capabilities, but the key features handled by the zone controller include:
- Space temperature monitoring.
- Temperature setpoint adjustments.

- VAV damper control.
- Airflow monitoring.
- Airflow minimum/maximum adjustments.
- Fan-assisted control sequencing.
- Electric reheat sequencing.
- Hot water valve control.
- Duct Temperature sensing.
- Warmup control sequencing.

The principal building automation control system programs the zone controllers through the data communications network. All monitoring points (such as space temperature) can be used for alarm and interactive controls applications. The basic programming of the terminal controller differs according to the manufacturer. Generally speaking, if the units are tied back into a data communications network, they all offer some type of control interface through a microcomputer such as an IBM PC.

Figure 12-1 shows a block diagram of a representative DDC/VAV terminal controller.

Interface Controls

Two of the important advantages of pneumatic controls (as opposed to electric, electronic, and DDC controls) are high reliability and low cost. However, the low accuracy of pneumatic controls has thrown doubt on their use in areas demanding more precise control. For other types of controls to utilize the advantages of pneumatic controls but retain the sophistication of DDC logic, electronic/pneumatic (E/P) transducers are required. Although E/P transducers are not new to the controls arena, some of the functions available on the new units will lend themselves well to retrofitting the DDC of existing pneumatic control systems.

The new line of combination controller boards with E/P transducers will accept pulse-width modulated (PWM) output or 4-20mA output from any source and convert that signal to a 0-18 psi output (in steps of less than 0.1 psi, in the case of PWM). In addition, the output pressure is monitored by the controller and is converted back

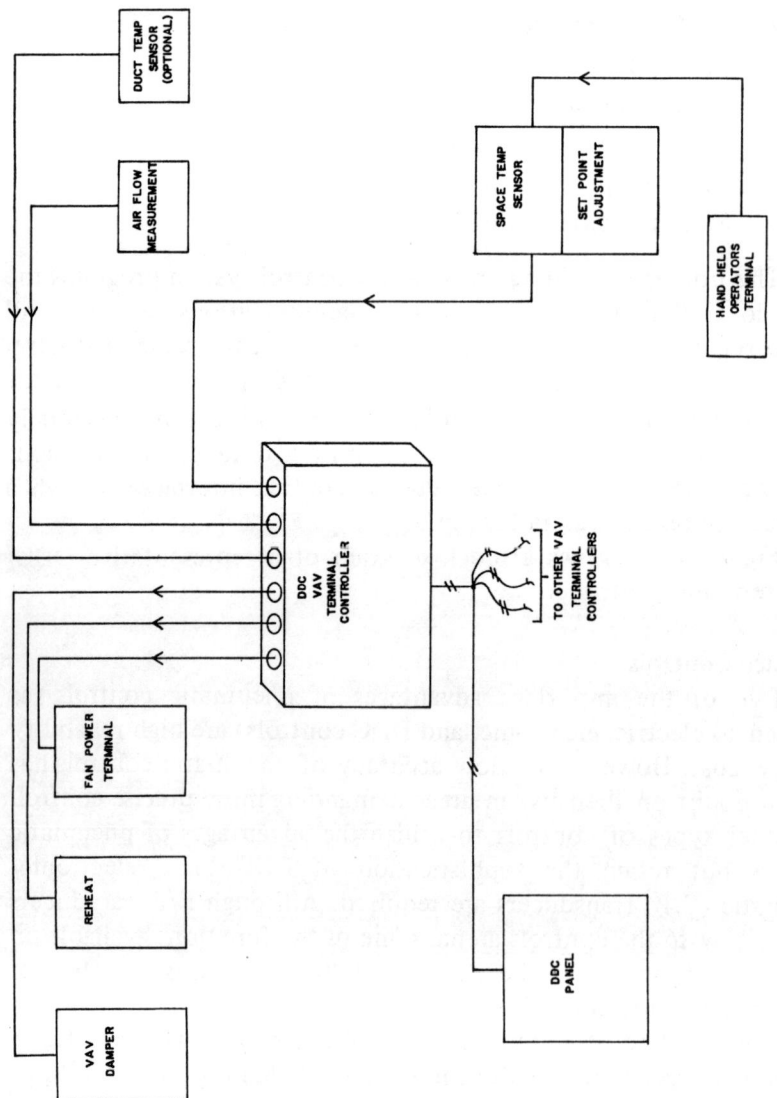

Figure 12-1. DDC/VAV terminal controller.

to a voltage feedback signal. This signal then goes back to the EMCS or DDC panel to read the actual output pressure being sent to the actuator. Failsafe features allow the transducer to transfer control to a standard pneumatic controller if the transducer stops receiving signals from the computer control system. A manual override switch also forces the controller to operate from the standard pneumatic controller. In retrofit jobs, using this type of output controller permits existing pneumatic control loops to be left in place as back-ups for the new DDC. A more detailed cost comparison is included in later sections of this chapter. But, for this single E/P transducer controller, the contractor price range is just over $300 each (uninstalled).

Figure 12-2 shows an example of a retrofit of DDC controls to a hot water converter system which is pneumatically controlled. In this case, the existing receiver controller would remain and become an input to the pneumatic port marked "original controls" on the controller card. The other two pneumatic inputs on the controller are the "Main" and "Branch" lines. Terminal inputs to the controller cards include the DDC electronic signal, pressure status feedback loop, and low voltage power supply. In this design, the temperature and pressure data are fed back to the DDC panel, whose software then calculates the proper control output to maintain the hot water discharge temperature. The electronic output of the DDC is converted to a pneumatic signal at the E/P transducer controller, and a pneumatic signal is sent to the steam valve actuators. The feedback loop from the E/P controller card monitors the output pressure, and sends a signal back to the DDC panel to be used in turning its PI control algorithm. If a failure condition is sensed by the controller, the controller automatically switches the signal output from the original receiver controller.

PERFORMANCE COMPARISONS OF HVAC CONTROLS

Each site has its own unique requirements which should be identified in order to determine what performance is expected from HVAC controls. While evaluating these requirements, decision makers should keep in mind the pros and cons discussed in the following paragraphs.

248　　　　　　　　　　　　　　　　　　　　OPTIMIZING HVAC SYSTEMS

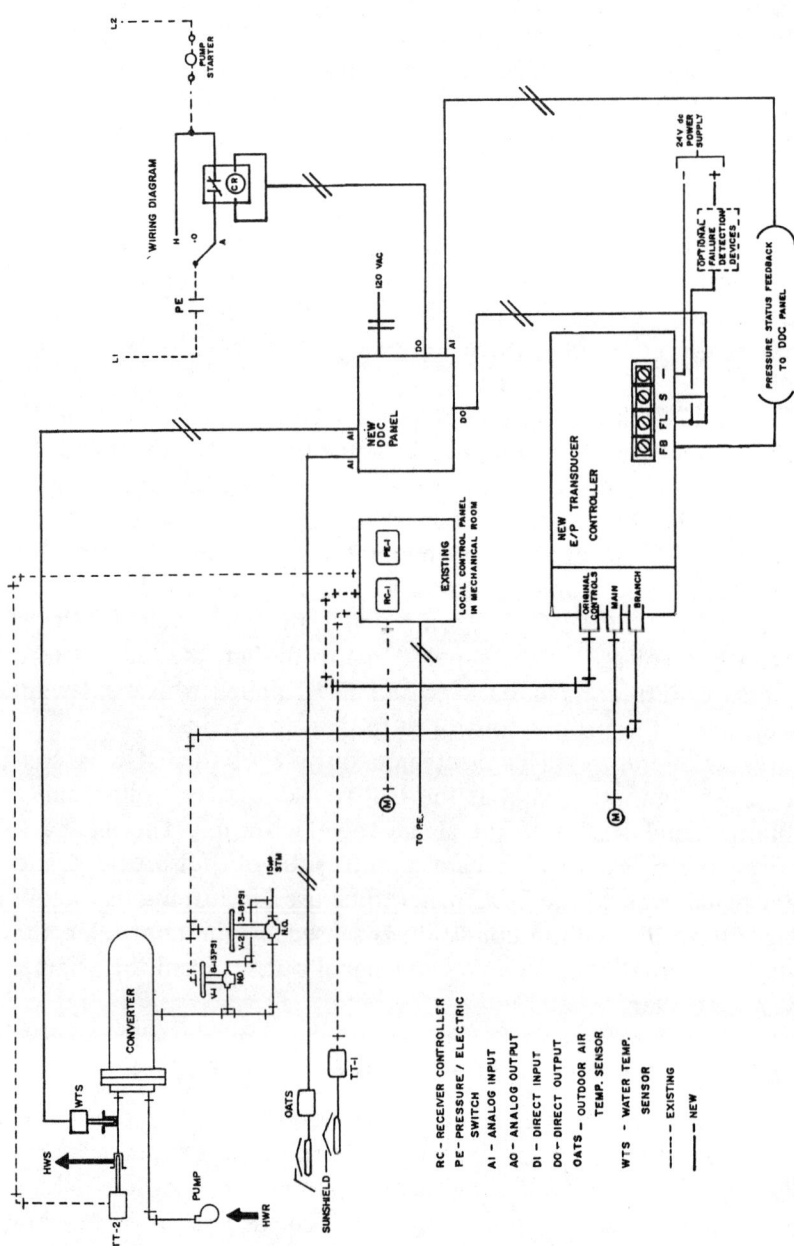

Figure 12-2. Interface controller.

Pneumatic Controls

For years, pneumatic controls have won the favor of engineers and technicians because they are simple to understand. In addition, pneumatic actuators are both inexpensive and reliable. This makes them very popular, for a wide range of applications, to control dampers and valves.

The disadvantages of pneumatics are threefold:

1. The instruments use very small orifices to modulate the air pressures. Unless the air source for the system is very clean, the controls will start to break down. The infiltration of oil or water into a pneumatic control system should be of great concern.

2. Pneumatic instruments are, generally speaking, the least accurate of all control systems. In recent published test results, accuracies of temperature transducers were ±4°F. The basic receiver control had an accuracy of ±2°F. The combination of a receiver controller with temperature transmitter had an accuracy of ±4°F[1].

3. Pneumatic controls have the largest number of moving parts in their controllers, which increases their tendency to go out of calibration quickly. The need to keep pneumatic instruments calibrated is an on-going process during the life of the controls.

Electric/Electronic Controls

In the HVAC market, basic electric/electronic controls of fan and terminal units are not as common as pneumatic controls. However, their use is sometimes advocated because of the advantages they offer. This category of controls provides an increase in accuracy over pneumatic controls. For example, the platinum resistance temperature detectors (RTD), which are used almost exclusively in electronic controls applications, can be built to provide tolerances of 0.5°F. They also never require site calibration, do have linear resistance to temperature profiles, are almost driftfree, and, most importantly, are inexpensive.[1]

The use of electronic controllers built with standard bridging networks and the use of high-quality RTDs will give very good control with limited calibration requirements. High-quality electronic controllers are available from both commercial and industrial control manufacturers. The best of these controllers can perform proportional (P), proportional-plus-integral (PI), or proportional-plus-integral-plus-derivative (PID) control functions for more precise control.

The greatest drawback to electric/electronic controls has been that, typically, electronic actuators were more expensive. In addition, electronic actuators are slower and typically require more maintenance than normal pneumatic actuators. For an application requiring a smaller actuator, such as a VAV terminal, an electronic actuator may cost the same or less than a pneumatic actuator. Some designers have approached this problem by using E/P transducers and pneumatic actuators to get the accuracy of electronic controls and still use cost-effective pneumatic actuators.

Direct Digital Controls

The use of direct digital control can take many application forms in HVAC controls. Not only can it be used to perform basic control sequences to maintain a constant static pressure on a VAV fan system, for example, but it can perform the full complement of energy management programs equally as well. This is where DDC will out-perform its standard pneumatic or electronic counterparts. As discussed under electronic controls, DDC offers control which is equally as accurate with electronic sensors such as RTDs. However, DDC falls into the same pitfall of using similar electronic actuators for standard interface. Many DDC manufacturers are now producing panels with direct E/P transducer output in the expectation that designers will want to use pneumatic actuators. Some DDC manufacturers even offer pneumatic inputs if a designer wishes to use a pneumatic temperature or pressure sensor.

The basic disadvantages of DDC against which skeptics have protested in the past are:

- Hardware reliability has not yet been proven.
- Programming DDC is too difficult.

- Too many functions are controlled by one processor.
- The cost of DDC is prohibitive.
- The increased control potential does not warrant the increased cost.

As mentioned previously, DDC manufacturers are starting to build third-generation DDC panels. Reliability problems are being designed out of the earlier versions of DDC systems.

The arguments used against DDC regarding high-point density control panels are also diminishing. With the further introduction of small DDC panels and DDC/VAV terminal controllers, the problem of mass control failure is greatly reduced. The cost of DDC is still a major concern for many purchasers of controls. However, with the new DDC/VAV terminal controllers, and with more factory-installed DDC equipment, labor and material costs should continue to drop.

The need for powerful control strategies may or may not be warranted for every application. Again, this gets back to the decision-making process: "How much control do I need?"

COST COMPARISON OF HVAC CONTROLS, 1986

Developing the final criteria for investing in a new HVAC control system, a comparison of cost for different alternatives is the most important factor. For each site, the way costs are entered into the decision-making process is different. At some sites, the initial costs of materials and installation are the only criteria used. For other facilities, a combination of first costs and life-cycle costs play an important role in choosing the best HVAC controls alternative.

Installation Costs of HVAC Controls

In order to better compare the costs of some of the different HVAC control methods, we estimated the cost of installing controls on a typical, medium-rise office building. This process was done for both a new installation and for a retrofit installation. This typical office building had the following features:
- Total square footage = 105,000.
- Eight stories high.

- HVAC type
  - VAV systems, with individual central fans on each floor.
  - Hot water reheat coils on exterior zones.
  - Minimum fixed outside air for ventilation.
- Central ventilation and exhaust fans.
- Central centrifugal chiller plant, with cooling tower strainer cycle.
- Central hot-water boiler plant.

The cost estimates included all the necessary hardware for the different types of installations. Cost estimates were adjusted and developed for six different geographic areas by applying local area labor rates, material cost factors, and local taxes. This was done to show the fluctuations in costs, as well as the need for individual analysis in selecting HVAC controls at each site.

Installed Costs, New Installation

For comparison, we developed three types of cost estimates for a new installation:

1. Basic pneumatic controls.
2. DDC system, using pneumatic actuators.
3. DDC system, using all electric/electronic actuators with new DDC/VAV terminal controllers.

The basic pneumatic control system used a standard thermostat and VAV boxes with hot-water reheat coils for zone controls. The central fan required discharge-air and fan-speed controls using single input receiver controllers. Miscellaneous hardware such as firestats, thermometers, and gauges were also included for the fan systems. Pneumatic controls and interlocks were included for the chiller, cooling tower, strainer system, boiler, building ventilation system, and associated auxiliaries. The necessary material and installation costs were also included for air compressors, air drying and filtering devices, and for pneumatic tubing.

We assumed that the DDC system with pneumatic actuators would use a completely electronic input and output control system, with pneumatic outputs for devices with actuators such as VAV dampers and valves. Overall, the DDC system was designed to provide all the basic temperature control functions as did the pneumatic controls previously described, but also using software to provide optimization and sequencing of control outputs. The cost estimate for this option was developed once with, and once without, a front-end personal computer (PC) for operator interface. This was done to illustrate the cost penalty for adding a PC for operator interface. The basic DDC temperature control system might not need the PC for normal control operation.

For the last DDC option, we developed cost estimates using all electronic sensing and actuator equipment. Included in this estimate were the DDC/VAV terminal controllers. Again, this DDC cost estimate is broken down into inclusion and exclusion of the PC front end to better compare the basic cost of this option with a strictly pneumatic control system.

Table 12-1 shows a breakdown of the three control options based on the descriptions above. These estimates are also broken down into six geographic areas of the United States to show the range of costs by area.

Table 12-1. Installed Cost of New HVAC Controls

City	Pneumatic Controls	DDC with Pneumatic Actuators		DDC with Electronic Actuators	
		Without PC	With PC	Without PC	With PC
Dallas	$105,000	$185,000	$211,000	$109,000	$135,000
Chicago	106,000	187,000	212,000	111,000	137,000
Denver	108,000	190,000	217,000	113,000	140,000
Los Angeles	109,000	194,000	219,584	118,000	144,000
Atlanta	103,000	181,000	206,000	106,000	131,000
New York	115,000	203,000	231,000	123,000	150,000

The results reveal a significant cost difference between a basic pneumatic control and a DDC system based around pneumatic actuators. The added control capability of DDC would have added almost 80 cents per square foot to the total construction costs.

However, for the DDC system with electronic actuators using the DDC/VAV terminal controller for basic temperature control, the costs are close to being competitive with those of basic pneumatic controls. To add a PC with software and graphic capability for operator interface, there is a $25,000 premium.

We generated separate cost estimates to examine the feasibility of retrofitting DDC controls in locations where there are existing controls. Table 12-2 shows the costs for replacing both a pneumatic and electronic control system with a DDC. The approximate $70,000 difference in cost between a new and a retrofit DDC with pneumatic actuator installation is due to the assumption that there are existing actuators available for valves and dampers. For DDC systems with a PC interface, the cost ranged from $900 per point for a pneumatic retrofit in New York, to $800 per point in Atlanta. The lowest cost overall was $550 per point for an electronic retrofit in Atlanta.

Table 12-2. Installed Cost of Retrofit HVAC Controls

City	Pneumatic Controls	DDC with Pneumatic Actuators		DDC with Electronic Actuators	
		Without PC	With PC	Without PC	With PC
Dallas	Base Case	$116,000	$141,000	$70,000	$96,000
Chicago		117,000	143,000	72,000	97,000
Denver		119,000	145,000	73,000	99,000
Los Angeles		122,000	148,000	76,000	102,000
Atlanta		113,000	138,000	67,000	93,000
New York		128,000	155,000	79,000	107,000

The costs for retrofitting DDC can be further reduced if the building has an existing supervisory EMCS system. Table 12-3 gives the estimated cost for using this type of retrofit DDC in the typical office building described. In this case, it was assumed not only that dampers and valves were available, but also that much of the wiring was already installed for interfacing to AHUs, chillers, boilers, and auxiliary equipment. This represents approximately $30,000 in savings between a retrofit DDC with a pneumatic actuator retrofit (see Table 12-2) and a retrofit DDC to an existing EMCS. The final cost per point for a DDC system with a PC interface would range from $500 to $700. On a square-foot basis, the average cost for a DDC interface to an existing electronic control system with an EMCS was found to be approximately 90 cents per square foot.

Table 12-3. Installed Cost of Retrofit HVAC Controls With Existing Supervisory EMCS

City	Pneumatic Controls	DDC with Pneumatic Actuators		DDC with Electronic Actuators	
		Without PC	With PC	Without PC	With PC
Dallas	Base	$84,000	$110,000	$64,000	$90,000
Chicago	Case	85,000	111,000	65,000	91,000
Denver		87,000	113,000	67,000	93,000
Los Angeles		89,000	114,000	68,000	94,000
Atlanta		82,000	108,000	63,000	88,000
New York		93,000	121,000	72,000	99,000

Building Energy Costs for Various HVAC Controls

To properly evaluate the various control options which have been discussed in this chapter, we performed an energy analysis. The purpose of this analysis was to estimate the building energy usage, based upon different control configurations.

The energy analysis involved the computer simulation of the same typical, medium-rise office building described in the cost estimate section. We assumed that the eight-story building was occupied only during a normal day shift, 5 days a week. Two basic simulations were carried out with the following control function modifications:

- Pneumatic Controls
    - Systems operate from 0500 hours to 1800 hours, Monday through Friday.
    - Heating setpoints: 71°F occupied, 55°F unoccupied.
    - Cooling setpoints: 75°F occupied, 85°F unoccupied.
    - Cooling-coil discharge temperature: 55°F.

- Direct Digital Controls (DDC)
    - Systems operate from 0600 hours to 1700 hours, Monday through Friday.
    - Heating setpoints: 68°F occupied, 55°F unoccupied.
    - Cooling setpoints: 78°F occupied, 85°F unoccupied.
    - Cooling-coil discharge temperature: 59°F.

The changes made in the two computer runs were to illustrate the difference in space temperature control, cooling-coil temperature control, and optimized start-stop timing. Because of the poor calibration in the pneumatic controls, cooling coils designed to be operating at 59°F may actually have operated at 55°F. This would have resulted in increased cooling and reheating. Likewise, the heating and cooling setpoints for the two simulations were used to represent the differences between local thermostats as opposed to temperatures maintained through remote control by DDC through software.

The simulations were performed on the DOE-2.B computer program, using TRY weather for six different geographical areas.[2] Monthly results were extracted from each run for electrical energy and demand, and for natural gas energy. This data, along with the actual utility rate structures for fuels in each geographical area, was used to determine the present energy cost per year. Table 12-4 lists the results of the energy simulations for both pneumatic control and DDC. The first two columns represent the annual electric and natural gas energy usage. Overall, for pneumatic and DDC controls, we found that the largest energy consumer by geographical region was Chicago, mainly because of its higher natural gas heating requirements. The minimum energy user in this analysis was the Los Angeles area. Figure 12-3 is a graphic representation of the comparison of the control types and cities.

Applying the local utility rates for energy and demand illustrated the significant differences, in annual dollar costs, between average city-to-city energy consumptions. These figures, listed in Table 12-4, range from a high in New York of $247,000 for pneumatic controls to only $52,000 for DDC used in Dallas. The bar chart in Figure 12-4 shows the annual energy costs in the six locations for both the pneumatic control and DDC simulation results. It is interesting to

Table 12-4. HVAC Controls Economic Summary

	Electric (MBtu/Yr)	N. Gas (MBtu/Yr)	Total (MBtu/Yr)	Energy Cost ($/Yr)	Life Cycle Cost ($/15 Yrs)	Installed Cost	Simple Payback	SIR
Dallas, Texas								
Pneumatic controls	5,550	1,794	7,344	64,369	665,080	105,000		
Direct digital controls	5,324	864	6,188	52,450	534,372	135,000		
Difference	226	930	1,156	11,919	130,708	30,000	3	4
Chicago, Illinois								
Pneumatic controls	4,949	3,861	8,810	147,822	1,371,013	106,000		
Direct digital controls	4,692	2,634	7,326	126,922	1,165,077	137,000		
Difference	257	1,227	1,484	20,900	205,936	31,000	1	7
Denver, Colorado								
Pneumatic controls	4,878	3,761	8,639	112,634	1,050,350	108,000		
Direct digital controls	4,627	2,468	7,095	94,136	868,646	140,000		
Difference	251	1,293	1,544	18,498	181,704	32,000	2	6
Los Angeles, California								
Pneumatic controls	5,033	1,684	6,717	186,291	1,862,320	109,000		
Direct digital controls	4,779	498	5,277	160,564	1,598,086	144,000		
Difference	254	1,186	1,440	25,727	264,234	35,000	1	8
Atlanta, Georgia								
Pneumatic controls	5,544	1,857	7,401	99,008	892,910	103,000		
Direct digital controls	5,320	897	6,217	82,567	730,786	131,000		
Difference	224	960	1,184	16,441	162,124	28,000	2	6
New York, New York								
Pneumatic controls	5,047	3,166	8,213	246,775	2,306,432	115,000		
Direct digital controls	4,795	1,850	6,645	211,691	1,965,275	150,000		
Difference	252	1,316	1,568	35,084	341,157	35,000	1	10

compare the results shown in Figures 12-3 and 12-4, where, for example, Los Angeles was found to have the lowest overall energy consumption but also the second-highest annual energy costs.

To complete the picture of the economic benefits for different control options, we performed a life-cycle cost analysis using the energy predictions from the computer simulations and the cost estimates developed in the previous sections. The life-cycle energy costs were determined by taking the annual energy costs and applying the Uniform Present Worth (UPW) factors for the different geographical regions over a 15-year lifetime at a 7% discount factor.[3] Again, New York had the highest overall, life-cycle energy cost ($2,000,000) for a typical office building with pneumatic controls. This contrasts with a minimum of only $500,000 for Dallas. Figure 12-5 shows the estimated, life-cycle energy costs for the six geographical locations.

We then calculated the cost premium for installing a DDC with electronic actuators and a PC interface (as opposed to using pneumatic controls in a new installation) along with the incremental energy savings to find the simple payback and savings-to-investment ratio (SIR). These calculations, presented in Table 12-4, showed that, for a new installation, the additional costs for DDC controls can be justified over a 15-year lifetime. In many cases, the economics indicated an SIR greater than six.

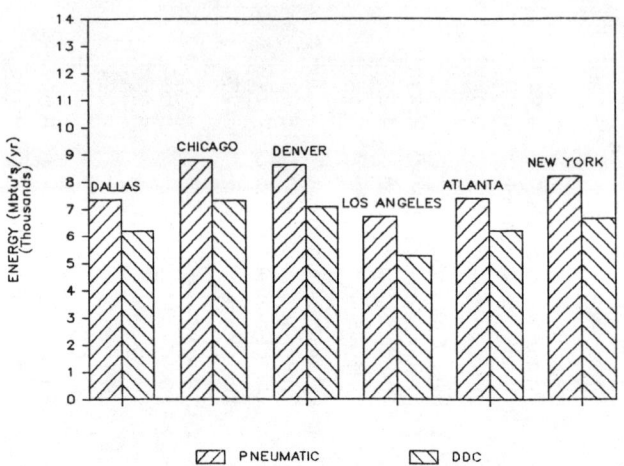

Figure 12-3. Annual energy comparison.

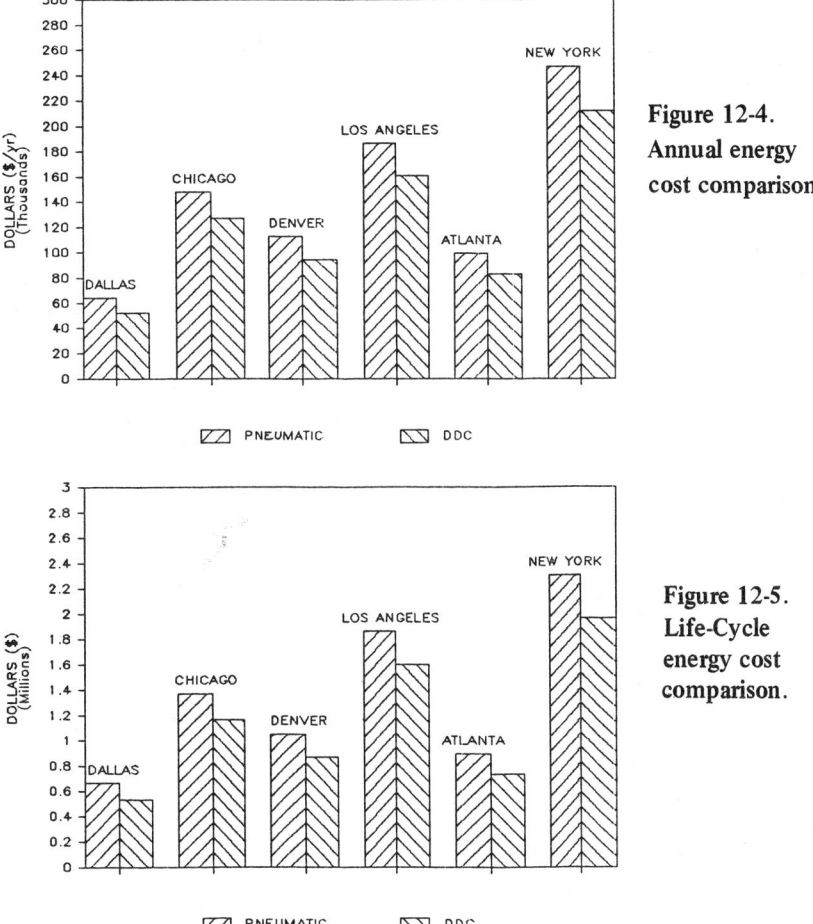

Figure 12-4. Annual energy cost comparison.

Figure 12-5. Life-Cycle energy cost comparison.

These numbers represent only hypothetical cases. However, we made conservative assumptions in simulating the typical office used for calculating the installed costs and energy costs for different control configurations. We concluded that the installation of a DDC would be favorable for similar type structures in most geographical areas of the United States.

SUMMARY

In the years to come, controls will be changing and developing at a pace equal to that of the microchip age. The current advances in microcomputer-based controls for chillers, boilers, HVAC, and teminal units will provide:

- Sophisticated control capability.
- Better reliability.
- Easier maintenance.

In designing a control system to be installed either in a new building or as a retrofit in an existing building, one should determine which criteria are important for that installation. Pneumatic controls are easy to use, less expensive, and have a high reliability rating, but they fall short on accuracy and flexibility. Electronic controls have much better accuracy, but their larger actuators are more expensive and less reliable. For applications requiring smaller actuators, such as VAV terminals, the cost for electronic actuators may be the same or less than the cost for pneumatic actuators. DDC provide the highest degree of control power, accuracy, and flexibility, but are more expensive for basic temperature control

The installed costs for pneumatic controls are still the most competitive for basic temperature control. The costs for DDC can vary widely, depending upon whether pneumatic or electronic actuators are used. The costs of DDC will also drop substantially for retrofit installations. Based on computer energy simulations, the savings available for DDC, as compared with those of basic pneumatic controls, should more than offset the premium costs for more powerful control features. The geographical differences in energy dollar costs vary widely and will greatly affect economic justification.

REFERENCES

[1] Hittle, D.C. and D.L. Johnson. "Energy Efficiency Through Standard Air Conditioning Control Systems." *Heating/Piping/Air Conditioning*, April 1986.

[2] U.S. Department of Energy, "DOE-2.1B." Building Energy Simulation Group, Energy and Environment Division, Lawrence Berkeley Laboratory, January 1983.

[3] Lippiatt, Wever, and Ruegg. "Energy Prices and Discount Factors for Life-Cycle Cost Analysis." United States Department of Commerce, National Bureau of Standards, November 1985.

13

Distributed Intelligence and Communication in Building Automation Systems

G.A. Bruns

INTRODUCTION

The cost of energy and the advent of reasonably priced microprocessors has resulted in some major changes in building control and communication systems. This chapter will address these changes and how they are affecting the way a building is operated.

Until recently, control of the heating and air conditioning systems in a commercial building has been by pneumatic control. From a control standpoint, pneumatic controls provide good, efficient and reliable control. The power provided by pneumatic actuators still makes pneumatic actuation the first choice.

The precipitous rise in the cost of energy has imposed some additional constraints on control systems, however, where pneumatic controls sometimes fall short. It is now necessary to monitor and coordinate all the control systems in a building to assure most efficient operation. In a large building with many HVAC systems, communication of information among the various control systems and to the manager of the facility becomes a significant factor. This is where the new microprocessor-based systems provide significant advantages.

HVAC functions are nearly all analog—meaning they are continuously variable. Temperature, humidity and pressure are all analog values. Output devices such as valves and dampers are also almost always analog—they can move to any position between full open and full closed. Pneumatic controls, some electric controls, and many electronic controls are inherently analog and, therefore, can directly monitor and control HVAC functions.

Microprocessors are strictly digital. That means they can only recognize a signal that is either full on or full off. They can count on-off pulses or take some sort of action in response to a signal going on or off, but they cannot deal with analog values such as temperature.

In early building management systems, there was just one processor, called the central processing unit or CPU. All the information from various sensors around the building was sent to this processor which monitored what was going on but usually did not perform any control functions other than to change the setpoint of a local control system or open or close a damper. It did not provide any direct control because of the possibility of significant delay in responding to changes in the controlled system and the serious problem that would result if the CPU failed—the whole building could go out of control.

SYSTEMS ARCHITECTURE

As microprocessors became lower in cost, it became practical to use many processors and distribute this intelligence around the building. This greatly increases the reliability of the system and makes it practical to directly control the HVAC, lighting and other systems by these distributed controllers. It also changed the requirements for communication among the processors and to a central. The architecture of the system has evolved into a tier system. A representation is shown in Figure 13-1.

Tier 1 includes the sensing and actuating functions typically found in commercial buildings.

Distributed Processors

The distributed processors are located in Tier 2. The DDC processor does the following:

- Receives information from sensors.
- Processes information according to program.
- Outputs signals to actuators.
- Stores information for future use.

Distributed Intelligence and Communication in Building Automation Systems 263

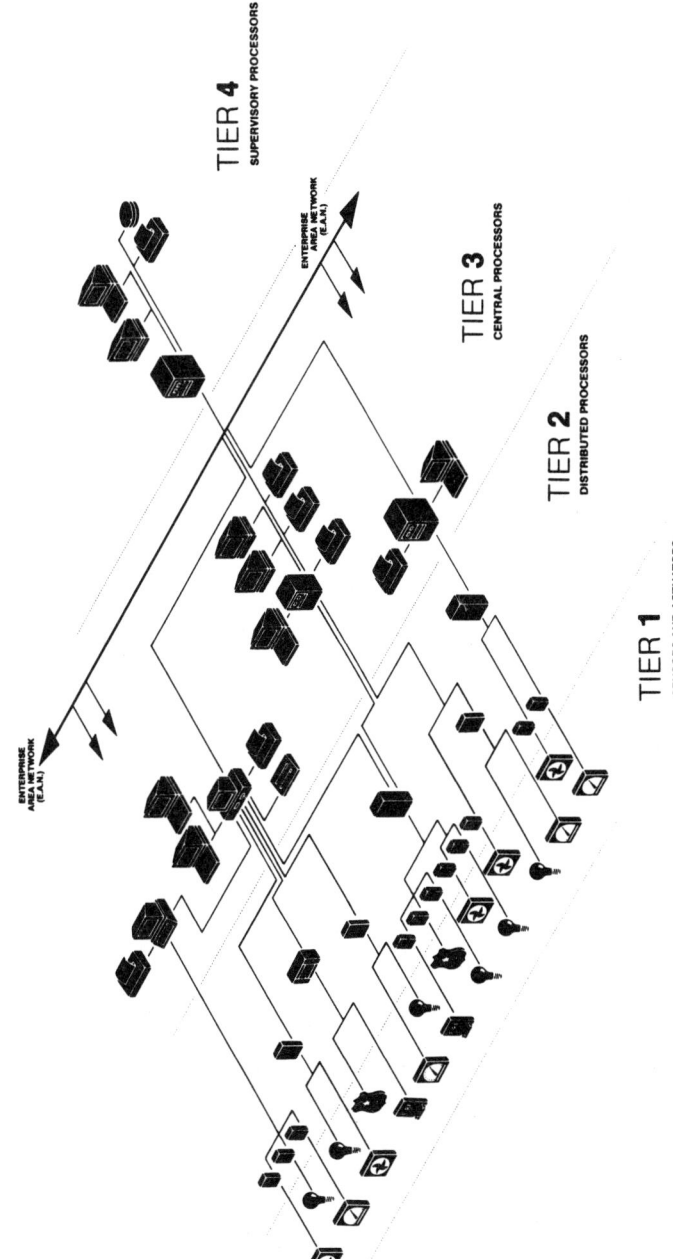

Figure 13-1. Systems Architecture.

- Provides control functions—can provide complicated functions more easily than pneumatic.
- Adaptive control.
- Has capability of communicating with other processors to send and receive information and commands.

The distributed DDC processors can be a single circuit board packaged in a well mounted panel (Figure 13-2) with features as follows:

— Full DDC control and EMS capability
— Battery backed real time clock
— Universal inputs
— LED operational indicators
— 16-bit microprocessor
— Operating system in EPROM
— Battery backed RAM for DDC and EMS parameters
— An accessory area for factory- or field-mounted transducers and relays
— Terminal trips for field connection to sensors and actuators

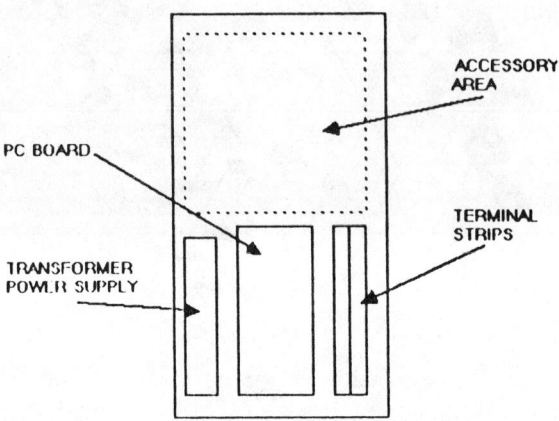

Figure 13-2. Distributed Processor Panel.

Standalone Distributed Processor: The distributed DDC processor can be applied in a standalone application. (See Figure 13-3.)

One processor typically serves one or two air handling units. This processor provides complete DDC control and EMS functions. A portable hand-held operator's terminal is available for functions such as:

- Display and change setpoint values
- Display and set application program data
- Enable/disable points, initiators and programs
- Display point summary and alarm summary

Data is accessed through descriptive menus for ease of use.

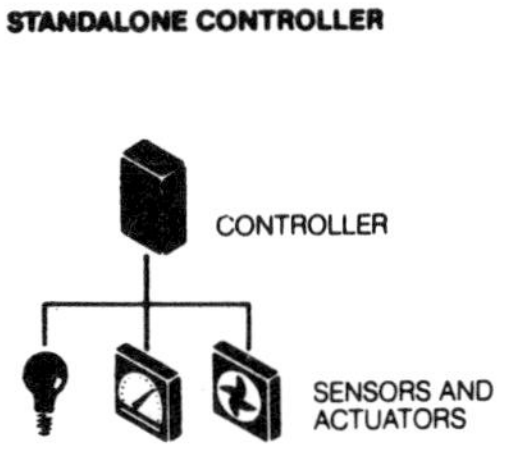

STANDALONE CONTROLLER

CONTROLLER

SENSORS AND ACTUATORS

Figure 13-3. Standalone Controller.

In larger systems, there is often information which must be sent from one controller to another and it is usually desirable to provide central access to all the information in the system. Most systems have provided this via a central processor or computer which relays information from one controller to another to gather information from all the controllers for display and printout and for total system energy management. This has been a very effective and efficient system but it still suffers from the major stortcoming of the earlier centralized system in that, if the central processor fails, communication between controllers is lost and the energy management programs in the central will not function.

Communication Among Controllers: The latest innovation in distributed processing is peer communication. Here the various controllers can communicate directly with each other without a central, and energy management programs can be resident in all the controllers.

In this system, multiple processors communicate via a Peer Network. With the Peer Network, every controller can communicate with every other controller as an equal or peer, with or without a central. There are several ways to allow the controllers access to the communication bus. An important feature must be that failed devices are automatically bypassed and, if the bus is cut, the individual segments must automatically re-establish communication and control.

Even with a fully loaded peer network, reporting of alarms and response to commands must occur within a maximum of a few seconds. Proper peer network communication protocol makes it possible to have distributed electric demand control without a central.

Peer communication makes it practical to share sensor data; e.g., all controllers have access to outdoor temperature from a sensor connected to one controller.

It is also feasible to have global event programs, again without a central.

The peer network should also permit downline and upline load of information from the central without taking controllers offline or stopping the global transfer of information.

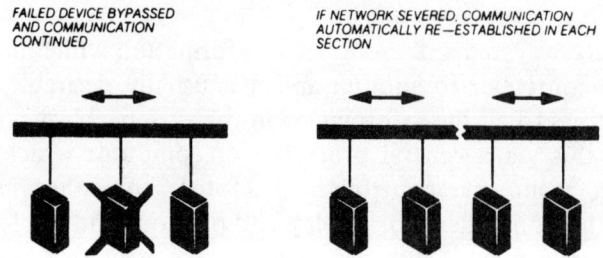

Figure 13-4. Peer Communication.

Central Processor

Central processors are located in Tier 3 of the systems architecture as shown in Figure 13-5. Video display terminals or personal computers provide a window to the system at this level.

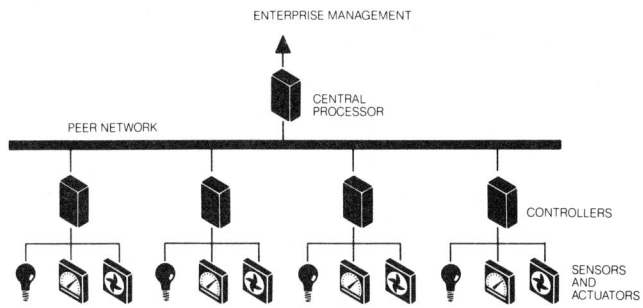

Figure 13-5. Central System.

Point data from multiple distributed control processors may be displayed in logical groups.

Alarms, operator responses, operator commands and returns to normal are automatically saved to disk. These can be retrieved later as a history log.

Critical operations can have temperatures or other analog values logged at regular intervals for later recall as an historical report.

Software Application Packages: Examples of software application packages include maintenance manager and energy auditor.

Maintenance manager automatically schedules and prints maintenance work orders and provides financial analysis of the costs associated with building maintenance. Maintenance schedules are generated based on calendar date, run time of equipment or an event. The system automatically prints work orders. A work order includes a description of the task, maintenance instructions, skill level, estimated time to complete and the necessary materials. Daily, monthly and yearly reports provide a summary of materials and skill costs and a summary of tasks that are completed or delinquent.

A simplified example of an energy auditor display is shown in Figure 13-6. Sensors for kilowatt hours (kW), killowatt demand and fuel flow rates provide data. The building owner inputs unit cost information. Energy auditor software then processes data to provide month-to-date and year-to-date reports. The reports can be expanded

BUILDING ENERGY AUDIT
REPORT FOR MARCH 1986

PARKWAY BUILDING
320,000 SQ FT

THIS MONTH

	1985	1986	% CHANGE
ELECTRICITY			
KWH	790,435	649,330	-18
KWH COST ($)	45,924	44,090	- 4
PEAK DEMAND (KW)	3525	3201	- 9
DEMAND COST ($)	14,276	16,645	17
NO. 2 OIL			
GALLONS	10,507	9141	-13
TOTAL COST ($)	11,663	8958	-23
MILLION BTU	1471	1280	-13
TOTAL ENERGY			
COST ($)	57,587	53,048	- 8
MILLION BTU	4168	3495	-16

Figure 13-6. Energy Auditor Option.

to include heating degree days, BTU's per square foot and other building efficiency indicators.

Other central processors may co-exist with, and share information with, the central processor just described. These central processors would be part of an Enterprise Area Network. An example of this is the hospitality automation system shown in Figure 13-7. This system provides total integrated control for the hotel business. It integrates the following environmental and control subsystems:

    Energy management
    Life safety monitoring/control
    Telecommunications
    Rooms management
    Property management

We will continue to see significant technology and software advances which will provide you with more usable information, even greater efficiency, more distributed intelligence and better/more reliable communication. You can expect to see intelligent controllers for individual terminal units such as the VAV box controllers. There will be more sophisticated communications systems. Fiber optics and power line carrier options in some systems and communication via telephone circuits within a building will soon be a reality. Digital communication systems will soon be standard so that telephones, computers, video and many other functions can use common lines.

In buying a new system or expanding an existing system, it is important to be aware of two important features—software and potential for expansion. Software is a very important feature because it is the determining factor in whether a system performs effectively or not. As the hardware becomes more generic—as in the expanding use of standard personal computers—the software makes the difference. A vendor's background in computers does not make him an expert in HVAC and energy management. Choose your software carefully.

It is also important that your system can be modified and/or expanded as your needs change and your building grows. Check your vendor to be sure that the vendor has accommodated older versions of a system in his new system. You do not want your system to become obsolete whenever a new system is announced.

270 OPTIMIZING HVAC SYSTEMS

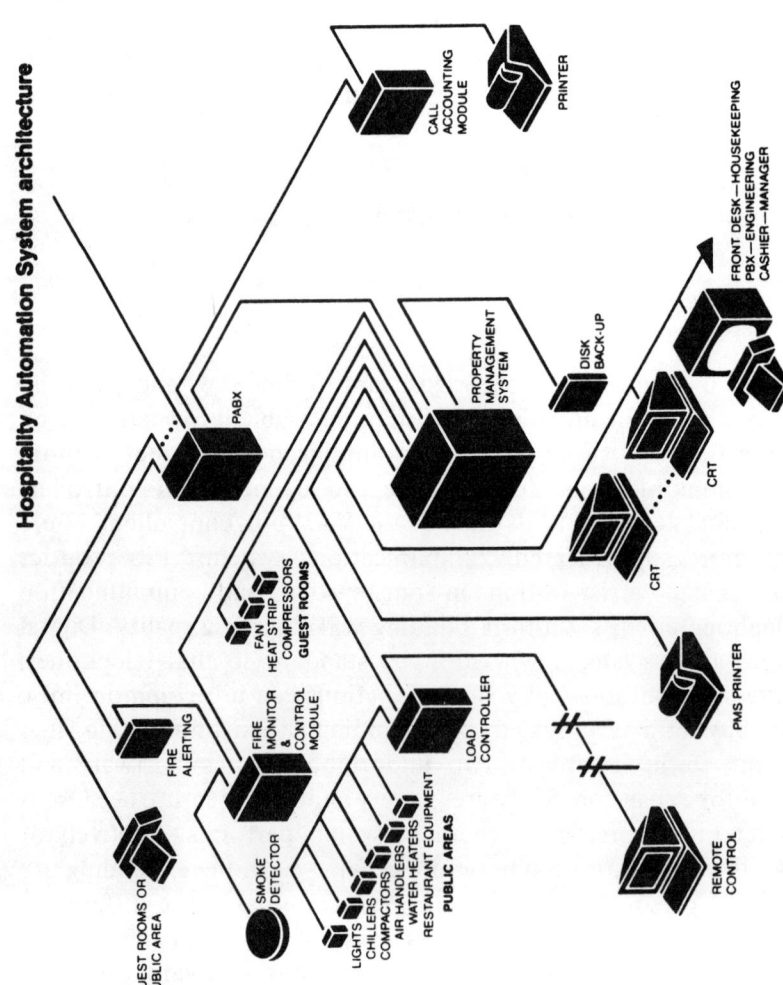

Figure 13-7. Hospitality Automation System.

SECTION VI
COOLING TOWER OPTIMIZATION

14

Cooling Tower Optimization — Innovations in Cooling Tower Heat Transfer

M.R. Lefevre

INTRODUCTION

A recent study, sponsored by the Environmental Protection Agency (EPA, see Ref. 27) shows that presently 51% of the water available in the United States is used for agricultural irrigation. By the year 2000 the projected number is 55%.

The Industry, including Power Plants, use approximately 39% today and by the year 2000 this will have to be reduced to 30% to satisfy other needs.

To achieve these objectives it is estimated that fresh water supplies to the industries will have to be recycled 17 times. For power plants the number is 7 times. These numbers are nationwide averages. For many regions the situation is already much more critical.

WET COOLING TOWERS — WATER LOSSES

Not too long ago wet cooling towers were considered the ultimate water conservation machine. Compared to a once-through cooling they only require 3 to 5% of the water otherwise needed with once-through cooling.

Today, providing this small amount of water needed to replace evaporation and other losses is becoming more and more challenging to the industry.

The only true losses in a wet tower are evaporation and blow-down. The blowdown is necessary to prevent the concentration of dissolved solids from increasing to the point where it may precipitate and scale up heat exchangers and the cooling tower fill, or reach unacceptable levels for other reasons.

The other losses, such a drift losses, windage through the air inlet and other leaks are not true losses; they can be subtracted from the necessary blowdown.

The best way to reduce water consumption is to reduce evaporation.

Evaporation. Wet Cooling Towers.
Influence of Ambient Conditions.

Evaporation depends on a large number of factors. Some are related to the air ambient conditions, and we have no control on those. They are:

Ambient Dry Bulb Air Temperature.

Ambient Air Relative Humidity.

Barometric Pressure.

Others are related to tower design and operating conditions and those can eventually be controlled by designers and users. Let's first see the influence of the non-controllable parameters on evaporation.

Table 14-1 gives the evaporation from a wet tower, as a function of ambient conditions.

As it can be seen there are substantial variations, but in practice this is beyond anyone's control, except if enough water storage capacity is available. In that case, it would be preferable to run the wet towers at night, with lower wet bulb air temperature and higher relative humidity.

As much as 20% water savings could be achieved in hot, dry climatic regions where wide variations of relative humidity and air temperature occur between day and night.

Table 14-1. Evaporation as a Function of Ambient Air Wet Bulb Temperature and Relative Humidity.

Wet Bulb		Relative Humidity %				
°F	°C	100	80	60	40	20
80		(1) .801	.828	.862	.909	.987
	26.67	(2) .345	.356	.371	.391	.425
60		(1) .702	.735	.775	.828	.906
	15.56	(2) .302	.316	.334	.356	.390
40		(1) .570	.602	.641	.690	.753
	4.44	(2) .245	.259	.276	.297	.324
20		(1) .412	.438	.467	.501	.540
	−6.67	(2) .177	.188	.201	.216	.233

Units: (1) lb water/1,000 Btu for English Units
(2) kg water/1,000 kJ for SI Units

Other conditions are held constant.

	English	SI
Barometric Pressure	29.92 in WG	1.053 Pa
Range	25 F	13.89 K
Approach	15 F	8.33 K
K Factor	0.6	0.6
Waterflow	Constant	
Heat Load	Constant	

Influence of Barometric Pressure on Evaporation

A tower located at 10,000 feet will evaporate approximately 4% more than its equipvalent at sea level. There is obviously nothing that can be done about it, but it is good to know. Table 14-2 illustrates the variation for typical conditions.

Table 14-2. Variation of Evaporation as a Function of Altitude

Altitude		Barometric Pressure		Evaporation	
ft	m	in Hg	ρa	lb/1000/Btu	kg/1000/kJ
0	0	29.92	1,050	0.862	0.371
500	152	29.06	1,020	0.865	0.372
1,000	305	28.07	985	0.869	0.374
2,500	762	27.07	950	0.872	0.375
3,500	1,067	26.07	915	0.876	0.377
10,000	3,050	20.06	704	0.897	0.386

All other parameters held constant.

Wet Bulb	80 F	26.67 C
Range	25 F	13.89 K
Approach	15 F	8.33 K
K Factor	0.6	0.6
RH	0.6	0.6
Waterflow	Constant	
Heat Load	Constant	

Influence of Design Parameter on Evaporation — Airflow

Let's see if we can decrease the evaporation by changing the airflow. For example, if we increase the airflow, both the air temperature leaving the tower and the absolute humidity decrease, but it does not mean reduced evaporation, unless the absolute humidity decreases more than the airflow was increased. Table 14-3 illustrates typical results.

Table 14-3. Influence of Air Flow and Relative Humidity on Evaporation.

K Factor	Air/Water Ratio	Units	Relative Humidity %				
			100	80	60	40	20
0.1	1.11	(1)	.775	.830	.891	.947	1.041
		(2)	.333	.357	.383	.408	.448
0.3	0.80	(1)	.785	.826	.877	.932	1.024
		(2)	.338	.355	.377	.401	.441
0.5	0.61	(1)	.796	.827	.865	.914	.997
		(2)	.343	.356	.372	.394	.429
0.7	0.48	(1)	.805	.831	.862	.904	.975
		(2)	.346	.357	.371	.389	.419

Units (1) lb water/1,000 Btu   for English Units
      (2) kg water/1,000 kJ   for SI Units

All other parameters held constant.

Range	25 F	13.89 K
Approach	15 F	8.33 K
Bar. Pressure	29.92 in WG	1,053 ρa
Water Flow	Constant	Constant
Bar. Pressure	29.92 in Hg	1,053 ρa

Unfortunately there is no general rule, there is no optimum airflow for which the evaporation is minimal.

For high ambient air relative humidities, there is a slight advantage to increase airflow. For low and average relative humidities, it is the other way around, lower airflows are preferable.

However, considering the variations of ambient relative humidity, it is always preferable, on average, to use as low an airflow as practical. This design for minimum airflow also coincides with wet towers most economical design and lowest fan energy requirements.

Influence of Range and Approach

The influence of range is minimal, at least within the practical limits of range variation.

Table 14-4 gives the variation of evaporation as a function of range at constant approach.

Table 14-4. Influence of Range on Evaporation

Range		Evaporation	
°F	°K	lb/1000 Btu	kg/1000 kJ
5	2.78	0.890	0.383
10	5.56	0.875	0.377
15	8.33	0.868	0.374
20	11.11	0.864	0.372
30	16.67	0.861	0.371
40	22.22	0.860	0.370

All other parameters held constant.

Wet Bulb	80 F	26.67 C
RH	0.6	0.6
Approach	15 F	8.33 K
Heat Load	Constant	
Waterflow	Inversely proportional to range	
Bar. Pressure	29.92 in Hg	1,053 pa

Table 14-5 gives the variation of evaporation as a function of approach at constant range.

Table 14-5. Influence of Approach on Evaporation.

Approach		Evaporation	
°F	°K	Btu/lb	kJ/kg
5	2.78	0.881	0.379
10	5.56	0.867	0.373
15	8.33	0.862	0.371
20	11.11	0.863	0.371
30	16.67	0.872	0.375

All other parameters are held constant.

Wet Bulb	80 F	26.67 C
Range	25 F	13.89 K
RH	0.6	0.6
K Factor	0.6	0.6
Waterflow	Constant	
Heat Load	Constant	
Bar. Pressure	29.92 in Hg	1,053 ρa

Table 14-6, the most significant one, gives the variation of evaporation as a function of range and approach combined, at constant hot water temperatures.

Table 14-6. Influence of Design Range and Design Approach on Evaporation at Constant Hot Water Temperature.

Range		Approach		Evaporation	
°F	°K	°F	°K	lb/1000 Btu	kg/1000 kJ
35	19.44	5	2.78	0.862	0.371
30	16.67	10	5.56	0.862	0.371
25	13.89	15	8.33	0.862	0.371
20	11.11	20	11.11	0.863	0.371
15	8.33	25	13.89	0.862	0.371
10	5.56	30	16.67	0.862	0.371
5	2.78	35	19.44	0.862	0.371

All other parameters are held constant.

Wet Bulb	80 F	26.67 C
RH	0.6	0.6
K Factor	0.6	0.6
Waterflow	Inversely proportional to range	
Heat Load	Constant	
Bar. Pressure	29.92 in Hg	1,053 pa

Note: It must be pointed out that this table refers to towers of entirely different design. For example, a tower for R = 35 F and A = 5 F is greatly different from a tower designed for R = 5 F and A = 35 F.

This is the most important one as it gives the variation at constant cooling capability (heat load).

It is sometimes a misconception to regard the cold water as illustrating the cooling capacity. As an example, the vacuum in a condenser is always a few degrees above the hot water temperature, regardless of range and approach. This is why we selected hot water temperature as our constant in Table 14-6.

Influence of Salinity

It is a well known fact that water with high dissolved solids exhibit a water vapor pressure lower than pure water at the same temperature.

Density and specific heat also vary with total dissolved solids (salinity).

Figure 14-1 illustrates the density variation as a function of salinity and temperature.

Figure 14-2 illustrates the specific heat variation with temperature and salinity. The influence on evaporation is significant: 8% less for 90,000 ppm salinity.

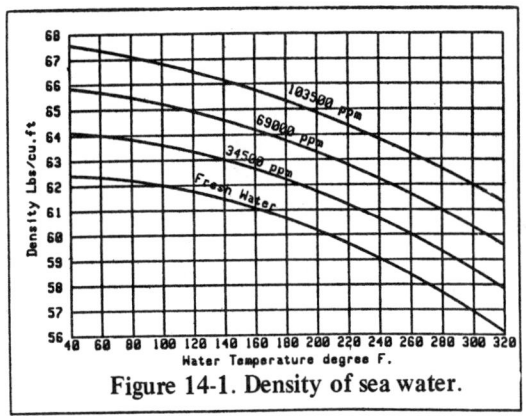

Figure 14-1. Density of sea water.

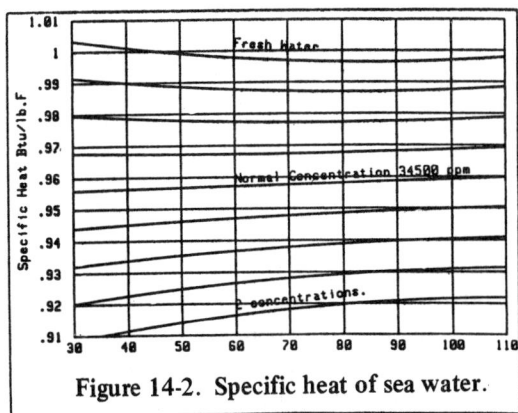

Figure 14-2. Specific heat of sea water.

Table 14-7 illustrates influence on evaporation.

Table 14-7. Influence of Salinity on Evaporation.

Salinity ppm	Density Ratio at 120 F	Vapor Pressure Ratio	Evaporation		% Mass	% Volume
			lb_{water}/1000 Btu	kg_{water}/1000 kJ		
0	1.000	1.000	.862	.371	100.0	100.0
30,000	1.022	.978	.857	.369	99.4	97.3
60,000	1.045	.956	.852	.367	98.8	94.5
90,000	1.067	.934	.847	.364	98.3	92.1

All other parameters are held constant.

Wet Bulb	80 F	26.67 C
Range	25 F	13.89 K
Approach	15 F	8.33 K
RH	0.6	0.6
K Factor	0.6	0.6
Bar. Pressure	29.92 in Hg	1,053 pa

This method may not be very practical, although it has been suggested to use highly concentrated solutions of lithium chloride to reduce and even eliminate evaporation entirely.

To the author's knowledge there is no large-scale industrial application of this principle.

Problems to be addressed with these high salt concentrations are scaling, corrosion, and strict drift control.

Wet Tower – Summary

There is not much that can be done in the design or operating mode of pure wet towers in order to reduce their evaporation losses. A few percent can be gained by using large storage reservoirs and avoiding to operate the tower during hot and dry ambient conditions.

At the design point of view it is recommended to use as low an airflow as practical. This requires, for a given cooling duty, highly efficient fill media. These are presently available in the form of modular film packings.

During part load operation it is recommended to operate at reduced airflow, using two-speed motors, variable speed, or variable pitch fans.

However, all of these precautions can only lead to minimal savings, at the most around 10% on a yearly average.

Crossflow versus Counterflow

Counterflow towers have, in theory, a uniform exit air wet bulb temperature.

On the other hand, crossflow towers exhibit a large variation of exit air wet bulb temperature. This is responsible for additional evaporation losses.

In addition, crossflow towers require more airflow to do the same cooling. Overall evaporation losses are very slightly higher.

The final difference is small, usually less than 1%. It is negligible at the water losses point of view.

This theoretically makes a difference at the environmental point of view because part of this additional evaporation recondenses by mixing and increases the density of fog leaving the tower.

In practice the additional airflow seems to compensate for that effect and it is hardly possible to tell the difference between a crossflow and counterflow plume.

COMBINATION OF WET AND DRY TOWERS

When there is not enough water available to provide the makeup for a conventional wet cooling tower, the only solution is to use "DRY" cooling to dissipate part of the heat load.

Separate Wet and Dry Towers

This arrangement is a simple and straightforward solution. One major good point in favor of such a solution is the use of two well known proven pieces of equipment: a wet cooling tower and a dry air cooler.

The air cooler usually uses fin tubes for heat transfer surfaces. Other surfaces, such as bare plastic tubes have appeared more recently on the market and look very promising for the future. The best arrangement of a dry and wet tower is in Figure 14-3.

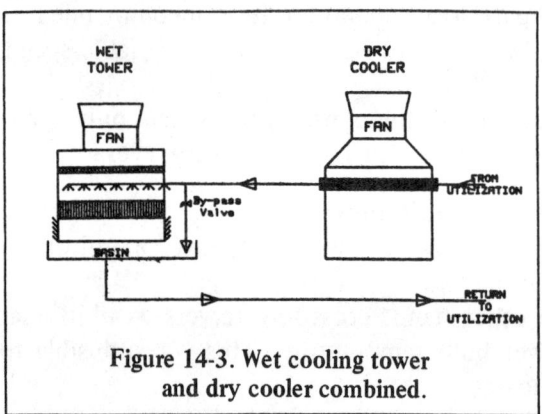

Figure 14-3. Wet cooling tower and dry cooler combined.

The water is cooled first in the DRY section because DRY cooling is much more expensive than WET cooling and this arrangement leads to the smallest DRY tower.

It must also be kept in mind that the DRY tower has a physical cooling limit equal to the dry bulb air temperature. The physical cooling limit for a wet tower is equal to the wet bulb air temperature.

The by-pass line shown on Figure 14-1 is a control device used to reduce evaporation. When the by-pass valve is fully opened the system operates as a pure DRY tower and the evaporation is nil. When the by-pass valve is fully closed maximum cooling is obtained.

To decrease the evaporation the trade-off is always a higher return water temperature from the system. Table 14-8 shows a numerical example.

Table 14-8. Variation of Temperatures as a Function of the Mode of Operation.

Dry Section	System Mode of Operation					
	Maximum Cooling		Operation			
Hot Water "IN" F, (C)	122	(50)	140	(60)	158	(70)
Dry Tower Δt F, (K)	18	(10)	27	(15)	36	(20)
Water "OUT" F, (C)	104	(40)	113	(45)	122	(50)
Approach F, (K)	18	(10)	27	(15)	36	(20)
Dry Bulb F, (C)	86	(30)	86	(30)	86	(30)
Wet Section						
Water "IN" F, (C)	104	(40)	114	(45)	122	(50)
Wet Tower Δt F, (K)	18	(10)	8		0	
Water "OUT" F, (C)	86	(30)	104	(40)	122	(50)
Approach F, (K)	8	(5)	27	(15)	NA	
Wet Bulb F, (C)	77	(25)	77	(25)	77	(25)

To decrease the evaporation it is necessary to increase the heat dissipation on the dry section.

To do this in practice, there are two methods and both consist of acting on the wet tower section.

By shutting off the wet tower entirely the system operates as a pure DRY tower and zero evaporation.

The penalty is an increase of 36 F, 122 F versus 86 F (20 C, 50 C versus 30 C) on the system returning temperature.

The first method, the easiest, consists of by-passing part of the waterflow to the WET PORTION. This method is dangerous during freezing conditions.

The second method, more elaborate, consists of reducing the airflow on the wet tower section. Variable speed motors or variable pitch angle fans are not necessary. In practice 2-speed motors on a multiple cell arrangement gives enough flexibility of operation, even during the worst winter operating mode.

The by-pass line is still necessary to keep the water from freezing in the basin during all DRY mode winter operation.

There are a few misconceptions about WET-DRY cooling which need to be clarified.

The first one is the amount of DRY surface needed to achieve a water conservation objective. For example, it is not necessary to design the DRY section to handle 50% of the heat dissipation at design conditions to save 50% water.

Particularly in the hottest, driest region of the USA, there are large variations of dry bulb air temperature between day and night. Even with a small make-up water storage capacity, it is easy to save water by reducing the cooling capacity of the wet section as explained before.

Table 14-9 shows an example for a 20 F variation in dry bulb. In this particular example the water savings can be doubled by a 20 F drop in ambient dry bulb temperature.

On an annual basis, the percentage water savings is much larger than the percentage at design point and important savings can be achieved, even with relatively small percentage of DRY cooling capacity.

Table 14-9. Example of Variation of Water Savings at Off Design Conditions.

	Design Max. Cooling		Off Design Minimum Evaporation	
Dry Section	(C)	(F)	(C)	(F)
Dry Bulb	40	104	30	86
Approach	5	9	10	18
Water "OUT"	45	113	40	104
Δt	5	9	10	18
Water "IN"	50	122	50	122
Wet Section				
Wet Bulb	28	82.4	25	77
Approach	7	12.6	25	77
Water "OUT"	35	95	35	95
Range	10	18	5	9
Water "IN"	45	113	40	104
% Dry Cooling and % Water Savings	33%		66%	

Extensive studies have been conducted for power plant applications and the results published (see reference list). How much water can be saved will depend on local conditions, how much storage capacity is available, performance penalty acceptable, heat load variation, water quality, and others.

Combined WET/DRY in the Same Towers

There are many possible combinations. The water circuit can be in series or parallel. The air circuit also can be either series or parallel.

On the air side the DRY section can either be traversed first by the ambient air or second, after traversing the wet section. This last arrangement will be favored when a visible plume reduction is also a secondary objective.

On the water side it is always better to feed the DRY section first to take advantage of the higher temperatures of the incoming water.

The choice of a series versus parallel arrangement for the airflow depends, among other things, on the ratio WET to DRY necessary to do the cooling. This is easy to understand by looking at the minimum airflow for the wet and dry tower sections separately.

The minimum airflow is dictated by the fact that the air cannot leave the tower warmer than the hot water coming in. For example, for the following conditions:

	(F)	(C)
Wet Bulb	70	21.1
Dry Bulb	90	32.2
Cold Water (Wet)	80	26.7
Range (Wet)	15	8.3
Cold Water (Dry)	95	35.0
Hot Water	110	43.3

A simple heat balance indicates:

$$\left(\frac{L}{G}\right)_{min} \quad \text{dry tower} = \frac{15}{0.24*20} = 3.13$$

$$\left(\frac{L}{G}\right)_{min} \quad \text{wet tower} = \frac{15}{63.32-34.09} = 0.51$$

(Note 63.32 = saturated air enthalpy at 95 F
  34.09 = saturated air enthalpy at 70 F)

(Units: Btu/lb.)

In this example, the air section needs about 6 times more air than the wet section and a parallel arrangement seems indicated.

Figures 14-4, 14-5, 14-6, 14-7, and 14-8 show schematically some of the configurations used.

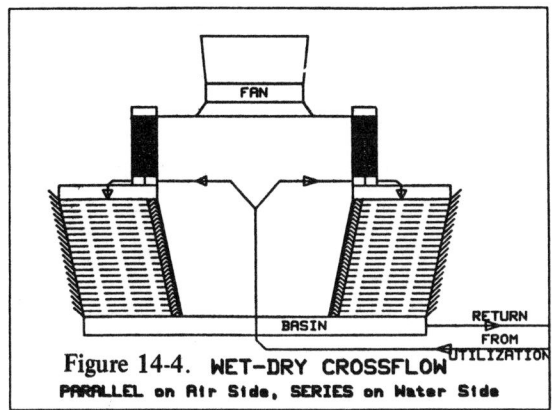

Figure 14-4. WET-DRY CROSSFLOW
PARALLEL on Air Side, SERIES on Water Side

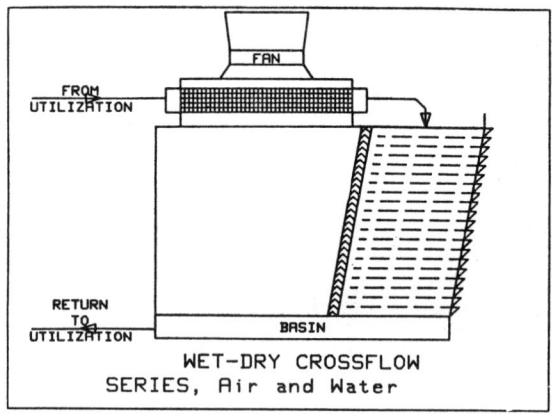

Figure 14-5.

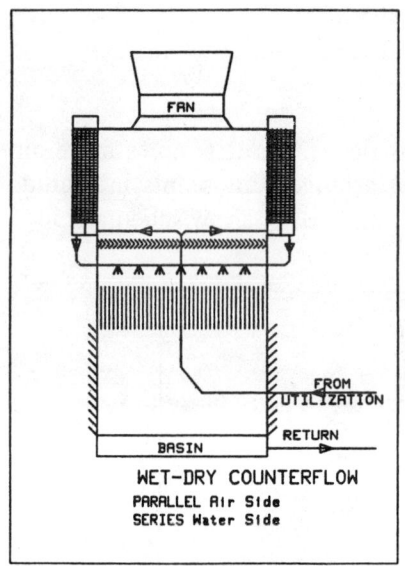

Figure 14-6.

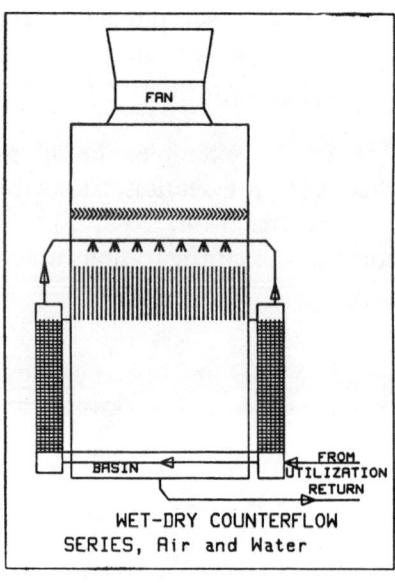

Figure 14-7.

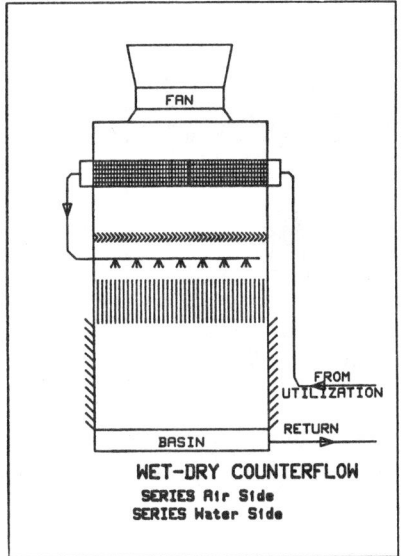

Figure 14-8.

Dry Tower With Peak Shaving

When very little water is available, one solution is to use a DRY tower but give it some help during the hot summer days.

The easiest way is to spray water directly on the fin tubes but this doesn't go without corrosion problems on the fins. In addition, it is not as efficient or desirable.

A more recent approach consists of using highly efficient modular evaporative pads in the air inlet of a dry air cooler. The water is a separate clean water circuit which is only used when additional cooling is required. When not in use the pads can be bypassed by air louvers (see Figure 14-9).

CONCLUSIONS

Presently evaporating water is still the most economical way to dissipate waste heat.

When insufficient water supply is available, there is little that can be done in the design of an evaporative tower to reduce evaporation.

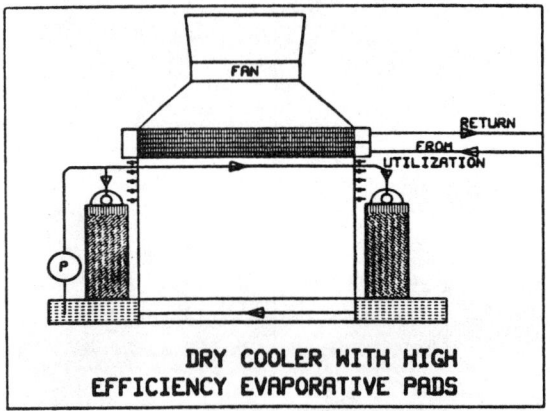

Figure 14-9.

Combination of wet towers and dry coolers, either as separate units or as combined system is the only presently known practical method to reduce evaporation.

REFERENCES

[1] A Survey of Capital Costs of Closed-Cycle Cooling Systems for Steam-Electric Power Plants. M.G. Monn, M.F. Rosenfeld, N.A. Blum. American Power Conference, April 1979.

[2] Waste Heat Management and Utilization Proceedings Conference, May 9-11, 1976, Miami Beach, Florida. University of Miami.

[3] EPA-600/7-78-157. Water Consumption and Costs For Various Steam Electric Power Plant Cooling Systems.

[4] Assessment of Cooling Water Supply in the United States. D.E. Peterson and J.C. Sonnichsen. American Power Conference, Chicago, Illinois, April 18-20, 1977.

[5] EPA-660/2-73-004. Nomographs for Thermal Pollution Control Systems.

[6] A Study of the Comparative Costs of Five Wet/Dry Cooling Tower Concepts. F.R. Zaloudek, R.T. Allemann, D.W. Faletti, B.M. Johnson, H.L. Parry, G.C. Smith, R.D. Tokarz and R.A. Walter. Sept. 1976.

[7] EPA-660/2-73-026. Technical and Economic Evaluations of Cooling Systems Blowdown Control Techn.

[8] A Dry Cooling Bed for Nuclear Power Station Condensers: Performances and Evaluation. C. Roma, University of Rome, 1974.

[9] Economics of High Back Pressure Turbines with Dry Cooling Systems. R.D. Mitchell, J.P. Rossie. American Power Conference, Chicago, Illinois. April 18-20, 1977.

[10] Report of the United States of America Dry and Dry/Wet Cooling Tower Delegation Visits to the Union of Soviet Socialists Republics. May 26 to June 7, 1975. Energy Research and Development Administration.

[11] Dry and Wet-Packing Tower Cooling Systems for Power Plant Application. M.W. Larinoff, L.L. Forster. ASME Annual Winter Meeting, December 1975.

[12] A Review and Assessment of Engineering Economic Studies of Dry Cooled Electrical Generating Plants. B.C. Fryer, March 1976.

[13] Combined Condenser Cooling Water Systems, M.B. Devereux. October 1968.

[14] The Disposal of Waste Heat from Power Plants. R.L. Herrick and R.M. Rosain. Cleveland Engineering Society Power Plants Division, March 4, 1975.

[15] Powered Spray Module, A New Concept in Evaporative Water Cooling Systems. Southeastern Electric Exchange.

[16] EPA-660/2-74-085. Effect of Geographical Variation on Performance of Recirculating Cooling Ponds.

[17] Proceedings of a Workshop on Power System Economics, Electric Power Research Institute and US Department of Energy, November 1978.

[18] Study of Dry-Type Cooling Towers and Their Application to Large Nuclear Power Plants. NTIS, U.S. Department of Commerce.

[19] Electric Power Generation With Dry-Type Cooling Systems. J.P. Rossie, E.A. Cecil, P.R. Cunningham, C.J. Steiert. American Power Conference, Chicago, Illinois, April 20-23, 1971.

[20] Wet/Dry Cooling Tower Study, Energy Research Development Administration, August 1975.

[21] ASME 75-WA/Pwr-10. Comparison of Different Combinations of Wet and Dry Cooling Towers.

[22] EPA-600/7-78-152. Optimization of Design Specifications for Large Dry Cooling Systems.

[23] Dry Cooler Study for Croydon Site of Philadelphia Electric Co. L.L. Haman.

[24] Zivi, S.M. and Brand, B.B., An Analysis of the Crossflow Cooling Tower, published in Refrigerating Engineering, 64, 1956, p. 31.

[25] Cooling Tower Institute, Cooling Tower Performance Curves.

[26] Kelly's Handbook of Crossflow Cooling Tower Performance, Neil W. Kelly.

[27] Assessment of Potential Environmental Problems Concerning Water Availability, U.S. Department of Commerce, June 1983.

15

Maximizing Cooling Tower Potential With State of The Art

R. Burger

ENERGY CONSERVATION IN REFRIGERATION

Enthalpy chart, Figure 15-1, illustrates that every 1°F colder water returned to the compressors in the operating range will reduce the power consumption of the system by 3½ percent. If the existing cooling tower can be retrofitted to produce 4°F colder water, this would result in a 14 percent less energy requirement. Reducing these to dollars and cents, a 4°F colder water times 3½ percent energy saved would equal a 14 percent savings. If the system requires, for example, $350,000 a year to power it, a 14 percent reduction in energy is translated into a $49,000 per year less cost of energy consumed. With projected energy increases practically every year, by maximizing the effectiveness of the cooling tower to produce a 4°F colder water, over $500,000 can be earned in a 10-year life of the equipment.

The three cell blow through tower at the North Texas State University was to cool 3,600 gallons per minute of water (1,200 GPM per cell) from entering the tower at 95°F, leaving at 85°F during a 78°F ambient Wet Bulb temperature. According to operating records of the Engineering Department, the system temperature averaged in the area of 100°F and the cooling tower discharge water was 89 to 90°F during the 78°F Wet Bulb design temperature requirement.

The rebuilding contract specified changing the inefficient corroded steel corrugated iron fill to highly efficient modern cellular

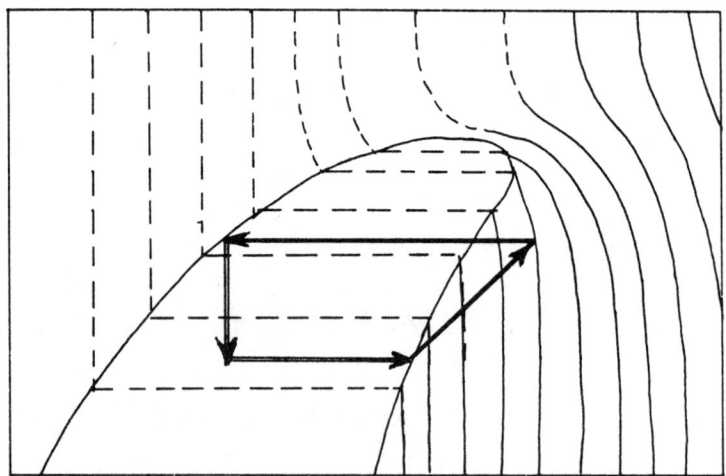

Figure 15-1. Enthalpy Chart indicates that cold water will save money in the refrigeration system.

film fill, Figure 15-2, the 388 small 1/8-inch-diameter orifice nozzles per cell, Figure 15-3, to the practically non-clogging large orifice (1-1/8-inch-diameter) ceramic square spray nozzle together with the associated 1½-inch-diameter old steel arms to new 3-inch-diameter PVC, Figure 15-4, and replace the corroded steel header with non-corroding PVC plus installing new low pressure drop drift eliminators.

Testing, in accordance with the Cooling Tower Institute ATC-105, indicated that the cooling tower was operating at the required levels of cooling the circulating water from entering at 95°F, discharging at 85°F during a 7°F approach to a 78°F Wet Bulb.

The original deficiency, caused partially by the scale within the piping system, clogging nozzles and rust creating dry spots in the fill, was cleared up by using the described state of art and the cost of $36,740 for this retrofit (which included sandblasting, priming and multiple coats of moisture cured urethane) brought this cooling tower up to desired operating parameters and the 4°F of colder water is calculated to save approximately $35,000 the first year which provides a rapid return of investment (ROI). In 10 years operating life this can approach $500,000 saved (or earned). Figure 15-5.

Figure 15-2. 388 small 1/8-inch-diameter orifices on 1½-inch pipe are prone to clog reducing cooling efficiency of tower.

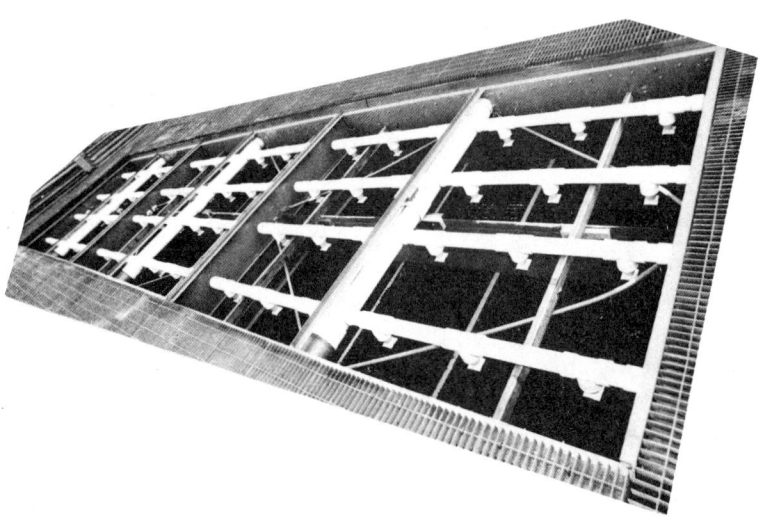

Figure 15-3. Retrofit of 36 large (1¼-inch-diameter) orifice square spray nonclogging nozzles of 3-inch-diameter PVC arms.

Figure 15-4. Sandblasted and coated steel with coal tar epoxy awaiting the installation of high efficiency heat transfer PVC fil cellular fill.

CASE STUDY: SAM HOUSTON COLISEUM

The 20-year-old cooling tower for the Sam Houston Coliseum which features the World's finest rodeos was in sad shape. The old-fashioned wood fill was collapsing, the wood water trough distribution system was leaking in many areas and clogging up in others, and many main structural supports were weakened to the point where a decision was made to replace the tower.

The Design Conditions of the new tower were to cool 6,000 gallons per minute of circulating water (3,000 gallons per cell) from entering the tower at 96°F, leaving at 86°F during an ambient Wet Bulb temperature of 80°F.

The existing tower under its best operating conditions was at least 5°F short of its requirement according to the Operating Engineer's log sheets.

> **REFRIGERANT ENTHALPY DIAGRAM**
>
> 1°F COLDER WATER = 3.5% LESS ENERGY
>
> REQUIRED TO OPERATE COMPRESSORS
>
> 4°F COLDER WATER × 3.5% = 14%
>
> LESS INPUT ENERGY NEEDED
>
> 14% × $350,000 = $49,000 PER YEAR
>
> OR $490,000 + EARNED FOR 10 YEARS

Figure 15-5. Cold water off the cooling tower saves money and provides rapid Return of Investment.

The internal elements of the tower incorporated the new state-of-the-art energy saving innovations including cellular fill heat transfer film fill which produces more cooling per cubic feet than conventional or other materials that are now being used, the same square spray ceramic nozzles were installed as shown in Figure 15-4, resulting in a very uniform distribution, Figure 15-6. Drift eliminators were changed to high efficiency, low pressure drop cellular plus redwood and stainless steel connectors used throughout.

The completed tower, Figure 15-7, was indeed energy efficient since the Cooling Tower Institute performed a certified test which produced 130 percent of capability, Figure 15-8.

This means that the new tower can cool the 6,000 gallons per minute of condenser water during the hot Houston 80°F Web Bulb temperature from entering the tower at 95°F, being cooled to leave the tower at 84°F. This 4°F Approach to the Wet Bulb, being 6°F colder than the old tower was producing, will save 18 percent of the energy costs to the refrigeration system which could produce approximately a $42,000-per-year savings in dollars to the City of Houston and help cool the cheering throngs as the cowboys do their thing.

Figure 15-6. The versatile square spray, nonclog nozzle and self-extinguishing high efficiency heat transfer surfaces are state of the art for cooling tower field erection and retrofit.

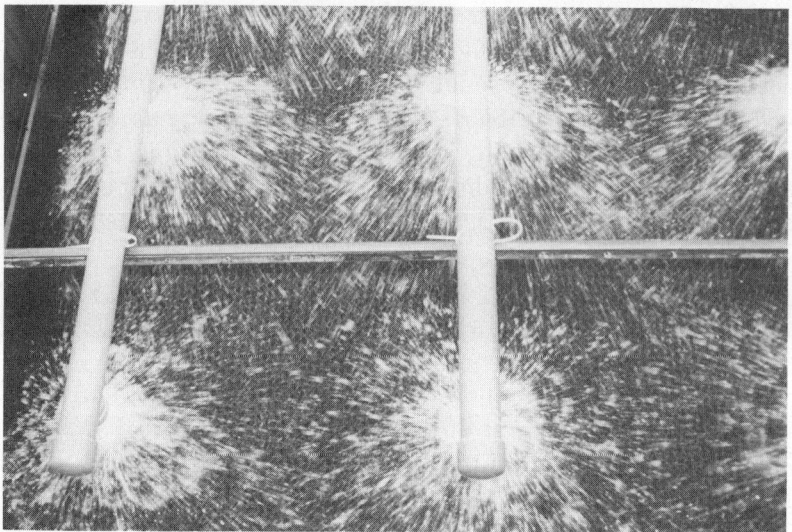

Figure 15-7. Square spray efficient pattern on cellular fill provides maximum cooling in towers.

Maximizing Cooling Tower Potential with State of the Art

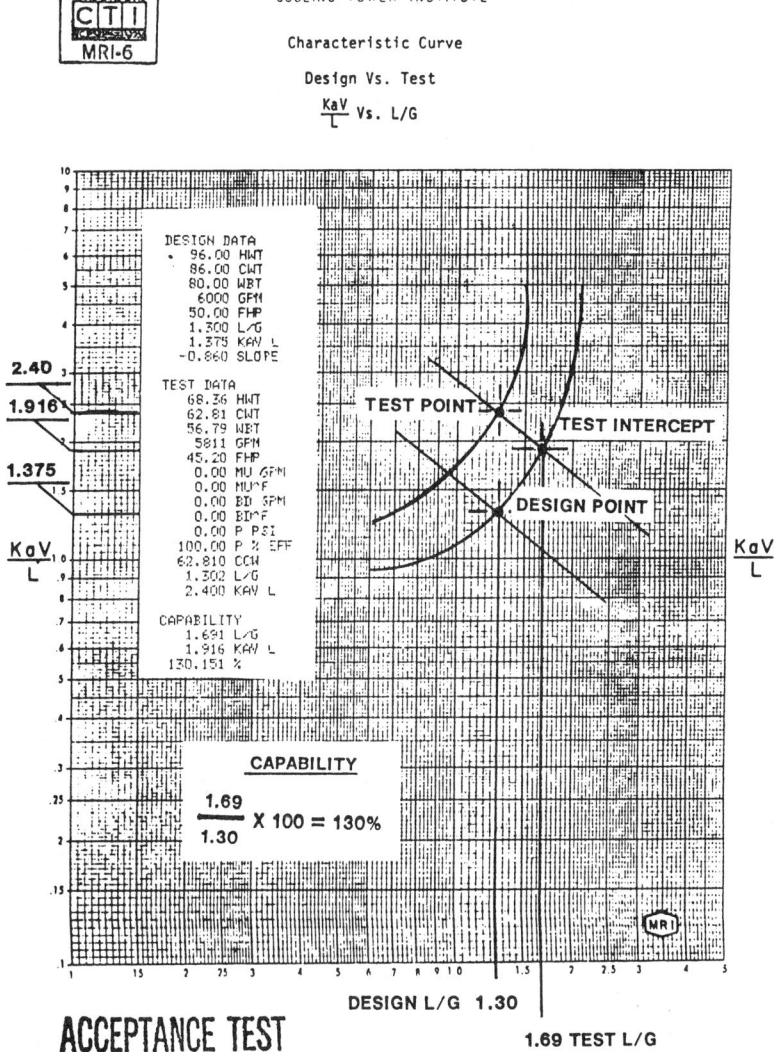

Figure 15-8. Results of Cooling Tower Institute ATC-105 Acceptance Test Code as performed by the impartial Mid-West Research Institute indicating a 130% capability which is 30% higher performance than the City of Houston paid for. This extra margin will provide a rapid return of investment for this installation.

CONCLUSION

While all cooling towers have this upgrading capability, they should be investigated along thermal engineering and state-of-the-art lines to determine the upgrading potential. A professional inspection and evaluation should be performed on cooling towers a minimum of 2 to 3 years apart, detailing the thermal, physical, structural, and mechanical equipment conditions with a report submitted, indicating recommendations, specifications and a budgetary estimate.

Figure 15-9. Field erected efficient cooling tower for the Houston Coliseum had all modern aspects of state of the art built into it including high efficiency heat transfer fill surfaces, uniform square spray nonclog water distribution pattern, cellular drift eliminators providing maximum drift control at low pressure drop, airfoil fiberglass epoxy propeller blades for maximum air movement, and velocity regained fiberglass fan stacks.

SECTION VII
HVAC
DESIGN
CONSIDERATIONS

16

Energy Effects of ASHRAE Standard 62-1981

C.E. Lundstrom

INTRODUCTION

During the last decade, escalating utility rates caused building owners and managers to take steps to lower their overall energy consumption. Among the measures energy managers took to reach energy optimization was the reduction of the quantity of outside air used for ventilation. As a result, there were increasing levels of indoor exposures to potential harmful products such as radon, asbestos fibers, unvented chemical vapors and combustion products, tobacco smoke, formaldehyde, and chlorinated organic chemicals. These developments have lead to the rewriting of the ventilation standard. This chapter will investigate the energy consequences of the new standard, ASHRAE Standard 62-1982.[1]

VENTILATION STANDARDS BACKGROUND

ASHRAE Standard 62-73,[2] "Standards for Natural and Mechanical Ventilation," was the first set of guidelines developed by ASHRAE which recommended volumetric air flow rates per person of "acceptable outdoor air" for ventilation purposes. In this standard, the quantities of air were specified as minimum and recommended values. Then, as energy awareness increased, ASHRAE Standard 90-75[3] used only the minimum ventilation values.

Today these standards are widely written into local and state building codes. Basically, this means that new and retrofit building HVAC designs need provide only 5 cfm of outdoor air per occupant.

After identifying the increased health risk problem with lower ventilation rates, ASHRAE re-evaluated and revised Standard 62-73 in order "to specify indoor air quality and minimum ventilation rates which will be acceptable to human occupants and will not impair health," using materials and methods which optimize efficiency of energy utilization. The new Standard 62-1981 was written and developed by an interdisciplinary group of engineers, physicians, chemists, and psychologists. Standard 62-1981 has a five-step ventilation rate procedure which prescribes:

1. The outdoor air quality acceptable for ventilation.
2. Outdoor air treatment when necessary.
3. Ventilation rates for residential, commercial, institutional and industrial spaces.
4. Criteria for reduction of outdoor air quantities when recirculated air is treated by contaminant removal equipment.
5. Criteria for variable ventilation when the air volume in the space can be used as a reservoir to dilute contaminants.

Higher ventilation rates are specified in areas where smoking is permitted, because tobacco smoke is one of the most difficult contaminants to control at the source.

Table 16-1 (taken from Table 3 of Standard 62-1981), shows the outdoor air ventilation requirements.

This means the minimum ventilation level recommended for an office building would increase from 5 cfm to 20 cfm of outside air per person if smoking is permitted, a fourfold increase. For building owners, increased ventilation could result in increased heating and cooling requirements (depending on location, existing internal and external loads, and methods chosen to meet Standard 62-1981).

ENERGY RESEARCH

This chapter documents computer simulation research done on the energy effects of ASHRAE Standard 62-1981 on an office structure. Before the new ventilation standard could be studied, it was necessary to choose a suitable office building to use as a representative computer model. Equitable Life allowed the Dravo Building in Denver, Colorado, to be used for the analysis. The building of rough-

Table 16-1. Outdoor Air Requirements for Ventilation in Commercial Facilities

Area	Smoking	Non-Smoking
	(cfm per person)	
Offices		
Office Space	20	5
Meeting & Waiting Spaces	35	7
Food & Beverage Service		
Dining Rooms	35	7
Bars & Lounges	50	10
Cafeterias, Fast Food	35	7
Retail Stores		
Sales Floor & Show Rooms	25	5
Malls & Arcades	10	5
Specialty Shops		
Barber and Beauty Shops	35	20
Florists	25	5
Sports & Amusement Facilities		
Ballrooms & Discos	35	7
Bowling (Seating Area)	35	7
Spectator Areas	35	7
Theatres		
Ticket Booth	20	5
Lobby, Lounge, Auditorium	35	7
Education Facilities		
Classrooms	25	5
Training Shops	35	7
Music Rooms	35	7

ly 150,000 square feet, built in 1977, had mechanical systems which represented an energy efficient design. The central plants in the building were two centrifugal chillers with heat recovery condensers which were used in conjunction with a 40,000-gallon water storage system and an electric boiler for space heating. The fan systems included two main VAV fans, each 70,000 cfm and 75 HP, which served the top seven floors. The fans were of the variable pitch vane axial type, controlled by a duct static pressure sensor.

Individual fan-powered VAV boxes with low-temperature HW reheat coils were used on the perimeters for cooling and heating requirements. Four pipe fan coils on the ground floor in individual retail and office spaces were used for space conditioning. The outdoor air was brought in through the main VAV fans. Modulating dampers on the outdoor air intake were controlled by pressure sensors to maintain the building at a slightly positive pressure. Time clocks and well calibrated local pneumatic controls maintained minimum runtimes on equipment and helped run equipment fairly efficiently.

COMPUTED DATA BASE

To effectively simulate the representative office building, past utility energy records and occupancy data were compiled. Electrical metering was done on building mechanical systems and the receptical and lighting systems for several months to identify equipment operating efficiencies and internal load profiles. Measurements were also made on the system supply and ventilation air flows. The resulting figures were compiled and averaged to represent an energy and building data base for the computer simulation.

The computer simulations were done on the building energy modeling program, ESP-II.[4] This software consists of four separate main programs which are as follows:
- Geographical Weather Data.
- Building Response Factors (Architectural Materials).
- Building Loads Analysis.
- Building Systems Analysis.

The weather program processes the National Oceanic and Atmospheric Administration (NOAA) weather tape to take the necessary hourly weather data for the energy programs. The weather program also adds solar position calculations and adjusts for latitude, longitude and elevation differences between the weather station location and building site location.

The response program computes the response factors used in simulating heating flow as a function of time across the boundry of the wall or roof.

The loads program computes the hourly heating and cooling loads for each space simulated in the building. This program takes into consideration the geometry of the space, walls and roofs of the space, external exposures, and hourly internal gains from people, equipment and process loads.

The systems program simulates the actual HVAC equipment hourly operation as it reacts to the hourly space loads generated by the loads program. Many secondary and unitary systems and central plant configurations are available. The program will total the energy consumption of the simulated building.

The first part of the research involved simulation of the base office building as it presently operates, using hourly weather tapes for the test reference year (TRY) for Denver, 1955. The computer-calculated monthly energy consumption and demand were then correlated with the actual metered and measured data for the modeled building.

After obtaining a base representative building, changes were made to the ventilation rates. First, the energy for the minimum ventilation rate (5 cfm per person) was modeled; second, the energy for the new proposed ventilation rate was modeled (20 cfm per person in general offices and 35 cfm per person in a conference room). The energy difference between the two runs was calculated. This data was then used, along with the utility rates charged by Public Services Company of Colorado, to calculate the hypothetical monthly and annual dollar increase caused by the proposed ventilation standard.

The same computer model was then used to anlayze the energy effects of ASHRAE 62-1981 in the following cities:

- Atlanta, Georgia
- Seattle, Washington
- Phoenix, Arizona
- Los Angeles, California
- Chicago, Illinois

- Dallas, Texas
- New York, New York

The TRY weather years used for each location were: Atlanta 1975, Seattle 1960, Phoenix 1951, Los Angeles 1973, Chicago 1974, Dallas (Ft. Worth 1975), New York 1951. By changing the weather data used in the computer simulation it was possible to predict the energy effect of ASHRAE 62-1981 in the different geographical areas, as it applies to an office structure.

Current local utility rates were used to calculate the monetary effects of the new standard.

COMPUTER SIMULATION RESULTS

The electrical energy for the base loads, fans, chillers, and boiler, plus the total monthly demand, were accumulated from the computer runs for each month for each geographical area studied.

Table 16-2 is a summary of the electrical energy, demand, and cost.

Of the eight areas studied, the most significant effects of the new standard were in Chicago. This was largely because Chicago has much higher heating requirements in the winter than the other areas. Energy usage in Los Angeles was reduced by the higher ventilation standard, basically because of lowered chiller consumption from an economizer effect. Figure 16-1 is a graph showing the overall energy effects of the two ventilation standards from the computer simulations.

Local electrical utility rates were applied to the energy and demand figures to calculate the monetary effects of ASHRAE 62-1981. The total dollar cost for energy varied widely between a low in Seattle of around $70,000 to a high in New York of approximately $480,000. On a percentage basis, Chicago again had the most significant increase in energy charges, with a difference of $25,800 between the possible energy charges from the two ventilation standards.

Figure 16-2 is a graph showing the overall dollar charges calculated from the computer-predicted energy and demand for the eight regions studied.

Table 16-2. Summary of Electrical Energy, Demand, and Cost.

	Atlanta	Seattle	Phoenix	Los Angeles	Chicago	Denver	Dallas	New York
ASHRAE 90-75 Minimum O.A.:								
Energy (kWh/Yr)	3,637,000	3,489,000	3,676,000	3,456,700	3,776,400	3,785,000	3,680,000	3,663,000
Demand (kW/Yr)	11,700	11,200	11,900	11,180	12,200	11,900	11,900	11,800
Cost ($/Yr)	338,200	70,900	222,200	302,500	286,200	212,100	206,600	482,200
ASHRAE 62-1981 Maximum O.A.:								
Energy (kWh/Yr)	3,786,000	3,691,000	3,751,000	3,448,200	4,127,300	4,040,000	3,823,000	3,910,000
Demand (kW/Yr)	12,400	11,800	12,200	11,240	13,200	12,400	12,700	12,600
Cost ($/Yr)	351,700	76,000	226,700	302,300	312,000	223,900	217,400	512,400
Comparison:								
Energy (kWh/Yr)	149,000	201,400	75,000	−7,500	350,900	25,500	143,000	247,000
Demand (kW/Yr)	700	600	300	60	1,000	500	800	800
Cost ($/Yr)	13,500	5,100	4,500	−200	25,800	11,700	10,700	30,200

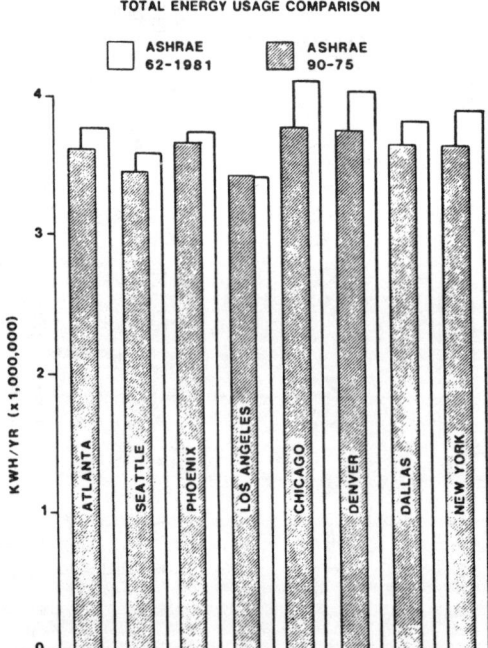

Figure 16-1. Total energy usage comparison.

The computer simulations for the eight geographical areas provided seemingly normal expected results. The accuracy in building simulations of energy usage is estimated to be 10-20 percent. The accuracy of this research, where the energy differential is calculated between two computer simulations for one variable such as outside air, should be in the 10% range. This research was done to obtain some representative figures for the magnitude of effects the new ventilation standard will have on building energy if it is widely accepted and implemented. To determine the best approach to meeting ASHRAE 62-1981, each individual application will need to be analyzed as new and retrofit designs of HVAC are developed.

The computer simulation results were verified by psychrometric analysis.

Energy Effects of ASHRAE Standard 62-1981

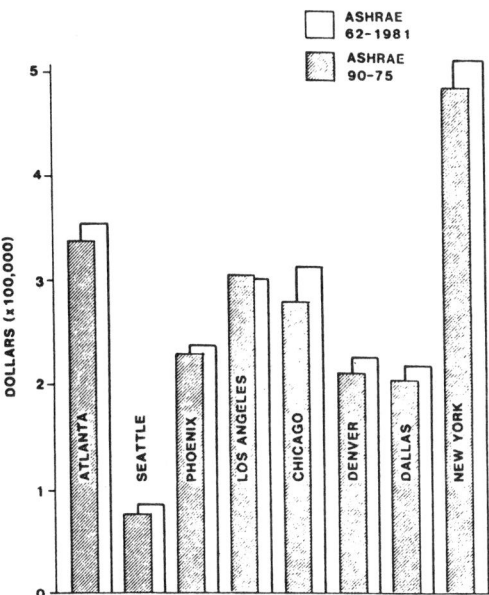

Figure 16-2. Energy cost comparison.

Figure 16-3 is a psychrometric chart showing a summer temperature condition of a central VAV fan system during a summer cooling mode. As shown in Figure 16-3, the amount of cooling required to go from point B (the mixed air condition going to the cooling coil) to point C depends on the percentage of outdoor air (point A) mixed with return air (point M). As more outside air is added to the mixed air stream, point B (mixed air) will move further toward point A (outside air). This represents a higher percentage of outdoor air to return air.

In this example, the difference in enthalpy (Δh) between the points represented by B will be the increased cooling required for ASHRAE 62-1981. The dashed lines running to points J, K, and L represent the part load sensible heat ratio (SHR) for the individual spaces. The solid line running to point M represents the full load SHR for the combination of all the spaces.

312 OPTIMIZING HVAC SYSTEMS

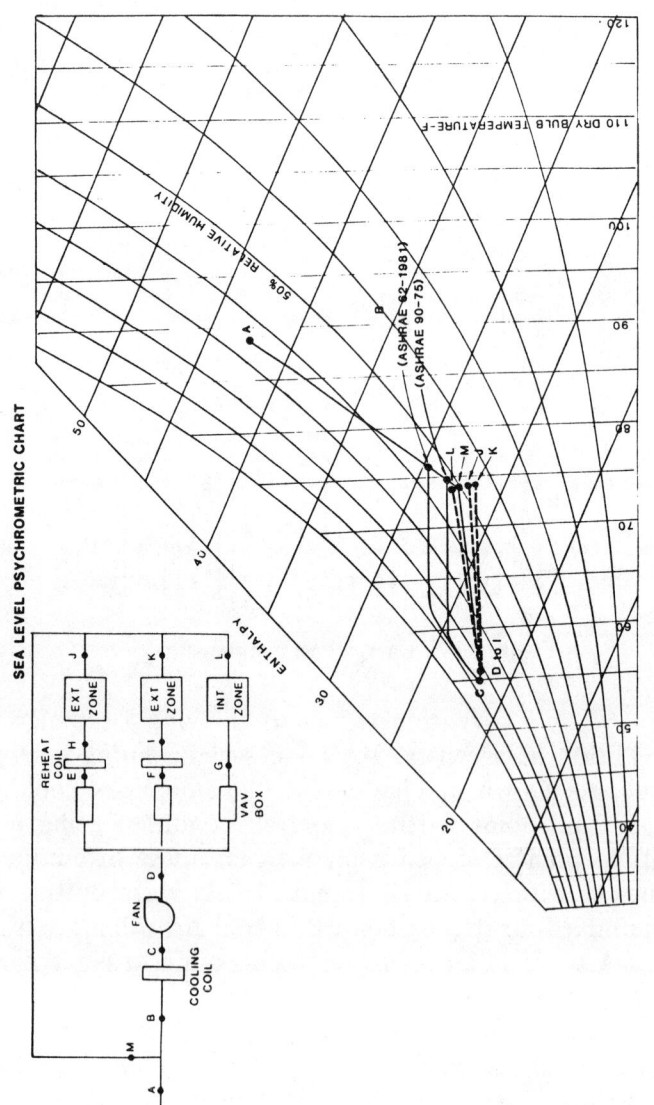

Figure 16-3. Summer conditions.

The same pshchrometric effects can also be determined for a winter period. In this case, increasing the ventilation rate will help reduce the cooling load on the entire building. Figure 16-4 shows the cooling decrease as represented by the enthalpy difference (ΔH) between the points represented by B. The overall energy effect will be less than the enthalpy difference; the representative building is served by a heat pump type system, which uses the waste heat from the cooling to heat exterior zones. Therefore, the less energy is used for cooling because of higher winter ventilation rates, the more backup electric heating will be required.

CONCLUSIONS

This study involved only one aspect of the ventilation requirements for ASHRAE 62-1981 for only one type of commercial building. The building used to model existing conditions represented an energy efficient mechanical system, which means that energy and dollar figures calculated by computer simulations for the study represent the minimum effects to be expected if ASHRAE 62-1981 is implemented. Another important fact which must be considered is that ASHRAE 62-1981 was written to provide alternative methods of having acceptable indoor air quality.

When ASHRAE 62-1981 was developed, a balance was struck between conserving energy and providing good health standards for building occupants.

According to computer simulations, ASHRAE 62-1981 could significantly affect energy consumption and costs if the standard is written into local and state building codes. Building owners and managers need to be aware of ASHRAE 62-1981, the requirement as it applies to different types of buildings, and the possible alternatives and approaches which can be taken to meet the ventilation requirement. In new and retrofit mechanical designs, outside air flows, filtering systems, and controls should be considered. Life cycle cost analysis should be done on the various design alternatives to determine the energy and dollar consequences of the ventilation standards.

314   OPTIMIZING HVAC SYSTEMS

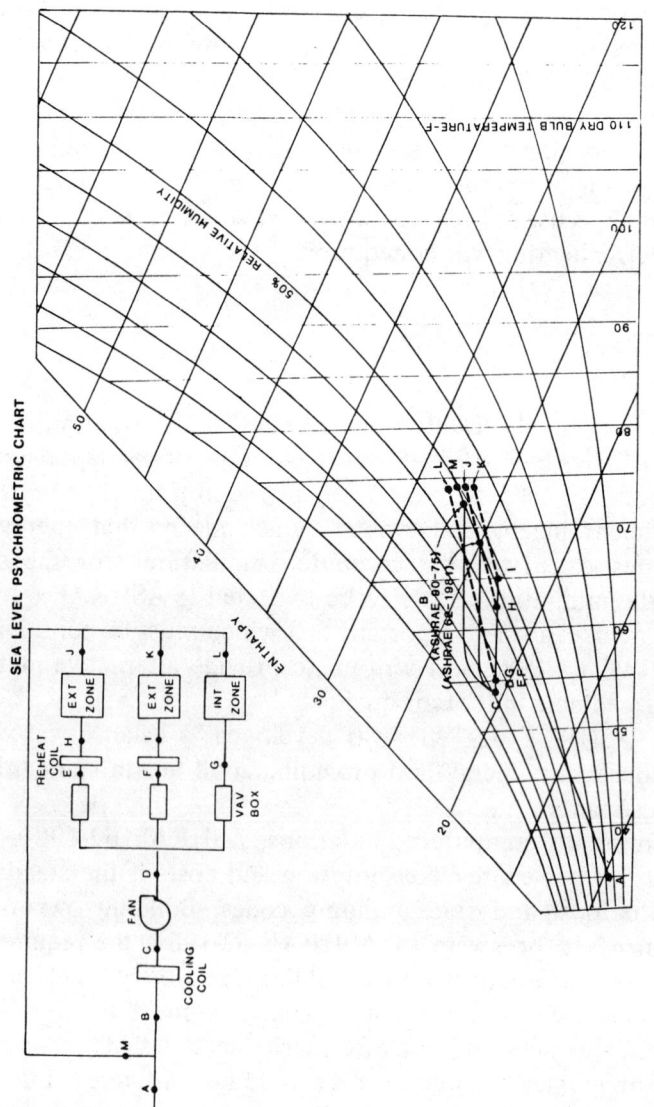

Figure 16-4. Winter conditions.

The overall energy and physiological effects of ASHRAE Standard 62-1981 require significantly more research in order to refine the standard to provide acceptable indoor conditions with a minimum of energy expenditure.

REFERENCES

[1] *ASHRAE 62-1981, Ventilation for Acceptable Indoor Air Quality;* American Society of Heating, Refrigeration, and Air-Conditioning Engineers, Inc.; Atlanta, GA, 1981.

[2] *ASHRAE 62-73, Standards for Natural and Mechanical Ventilation,* American Society of Heating, Refrigerating, and Air-Conditioning Engineers, Inc.; Atlanta, GA, 1973.

[3] *ASHRAE 90-75, Energy Conservation in New Building Design,* American Society of Heating, Refrigerating, and Air-Conditioning Engineers, Inc.; Atlanta, GA, 1973, 1975.

[4] *ESP-II Computer Program,* Automated Procedures for Engineering Consultants, Inc.; Dayton, Ohio, 1981.

17

Can Absorption Chillers Beat Electric Centrifugal Chillers?

W.F. Kirsner

INTRODUCTION

A recent issue of *Energy Decisions* published by Gas Energy, Inc. asked the question in one of their headlines—"Is Electric Cooling Becoming Obsolete?"

A cost comparison of chiller operation performed by an Atlanta area vendor of high efficiency absorption and electric centrifugal chillers purported to answer this question in Georgia. The answer: 100% direct fired gas air conditioning achieves a substantial operational savings over electric centrifugal chillers of the same size— enough to pay for the added cost of the absorption chillers in 2 years and 8 months!

This chapter also purports to answer the question posed above. It presents a generalized analytical technique which can be used to compare the run cost of gas or steam fired chillers with that of an electric centrifugal chiller. What's more, it provides a definitive answer to the question posed in the title of this chapter for typical buildings in Georgia, billed under the Georgia Power PL-6 rate schedule.

ABSORPTION CHILLER RUN COSTS

A typical single-stage steam absorber requires 18 pounds of steam to produce a ton-hour of refrigeration. At 1,000 Btu per pound of steam, .8 boiler efficiency, .9 distribution efficiency (to deliver the steam to the absorber) and 40 cents/therm of gas[1] (which corresponds to the Interruptible Rate for "Summer Air Conditioning" charged by the Atlanta Gaslight Company), the cost per ton-hour to operate a single-stage absorber is 10 cents per ton-hour.

$$\frac{18 \text{ \# steam}}{\text{ton-hr}} \times \frac{1000 \text{ Btu}}{\text{\# steam}} \times \frac{1}{.8 \text{ boiler eff.}} \times \frac{1}{.9 \text{ dist. eff.}} \times \frac{40¢}{\text{therm}} = \frac{10¢}{\text{ton-hour}}$$

A second more efficient type of absorption chiller is a two-stage or "double effect" absorber. Most units now in use were manufactured by the TRANE Company. They typically use about 12 pounds of steam to generate a ton-hour of cooling. Their run cost is 6.7 cents/ton-hour.

$$\frac{12 \text{\# steam}}{\text{ton-hr}} \times \frac{1000 \text{ Btu}}{\text{\# steam}} \times \frac{1}{.8} \times \frac{1}{.9} \times \frac{40¢}{\text{therm}} = \frac{6.7¢}{\text{ton-hr}}$$

However, Hitachi recently introduced a more efficient double effect steam absorber which uses only 10.6 pounds of steam to produce a ton-hour; its run cost would be only 5.9 cents/ton-hour.

$$\frac{10.6 \text{\# steam}}{\text{ton-hr}} \times \frac{1000 \text{ Btu}}{\text{\# steam}} \times \frac{1}{.8} \times \frac{1}{.9} \times \frac{40¢}{\text{therm}} = \frac{3.9¢}{\text{ton-hr}}$$

The most efficient absorbers commercially available are the direct-fired variety made by several Japanese firms. They have a C.O.P. of approximately 1.0, indicating that they require about 12,000 Btu of energy input to produce a ton-hour of cooling. But, there's a catch here; a footnote in the manufacturer's catalog explains that the input energy is expressed in terms of the "lower heating value" of the fuel being used.

Some comparisons of the lower and higher heating values for various fuels are shown below:

LHV versus HHV for FUELS

	L.H.V.	H.H.V.	LHV/HHV
Natural Gas	930 Btu/c.f.	1,030 Btu/c.f.	90%
#2 Fuel Oil	131,500 Btu/gal	138,500 Btu/gal	95%
#6 Fuel Oil	142,000 Btu/gal	150,500 Btu/gal	94%
Coal (Bitum)	13,600 Btu/lb.	14,100 Btu/lb.	96.5%

For all fuels, the lower heating value is less than the higher heating value. The difference is in the Btu's which are contained in the heat of vaporization of water vapor formed during the combustion process. If the gaseous products of the combustion process can be condensed, then one can re-capture the higher heating value of a fuel as useful energy. But, since condensation of corrosive combustion gases is normally not possible, then the lower heating value of a fuel is generally the maximum energy that can be extracted from it.

In a direct fired absorber, the combustion gases cannot be condensed, hence only the LHV of the input fuel can be used. BUT the Gas Company charges for natural gas based on its higher heating value. In a sense, the absorber manufacturer and the Gas Company look at the same cubic foot of gas and, because of their differing perspective, one sees 930 Btu there, while the other sees over 1,000 Btu.

The upshot is that while a direct fired absorber manufacturer may say that you need 12,000 Btu's of natural gas energy to produce a ton-hour of refrigeration, you will have to buy 13,290 Btu from the Gas Company:

Thus the effective C.O.P. for a direct fired absorber using natural gas based in the HIGHER HEATING VALUE of the fuel is:

$$\text{C.O.P. (HHV)} - \frac{1 \text{ ton-hour cooling}}{13,290 \text{ Btu HHU}} - .87$$

The run cost of the direct fired absorber, then, is:

$$\frac{13{,}290 \text{ Btu HHV}}{1 \text{ ton-hour}} \times \frac{40¢}{\text{therm}} = \frac{5.3¢}{\text{ton-hour}}$$

Now, let's compare the run cost of a high efficiency electric centrifugal chiller requiring .6 kW/ton of electricity to produce cooling.

ELECTRIC CENTRIFUGAL RUN COST

Electric energy for a good-sized building billed under the Georgia Power PL-6 Rate Schedule will typically cost about 2½ cents/kWh.[2] Thus, at .6 kW/ton, the cost to operate a high efficiency electrical chiller is about 1.5 cents/ton-hour + Demand.

$$\frac{.6 \text{ kW}}{\text{ton}} \times 2.5¢/\text{kWh} - \frac{1.5¢}{\text{ton-hour}} + \text{Demand}$$

The demand charge is an additional monthly charge for electricity based on the electric power drawn during that month or any of the previous 11 months. We need to know the amount of the demand charge if we are to complete our comparison.

First let me start by stating a problem ...

Demand is a power charge, not an energy charge. Thus it cannot precisely be quantified on a ton-hour basis, which is the method of comparison we are using in this analysis. The best we can do is to look at the "average" charge for billing demand for a *specific building's* Chilled Water Plant over the period of a year and then divide by the annual ton-hours produced. This will yield an average **demand charge/ton-hour** for that building.

Let's look at an example:

Grady Memorial Hospital. The 7th floor Chilled Water Plant at Grady Memorial Hospital in Atlanta, Georgia, produced 2,150 tons at peak load and 6,143,000 ton-hours annually. The annual billing demand charge contributed by the electric centrifugal chillers at Grady is calculated as follows:

Can Absorption Chillers Beat Electric Centrifugal Chillers?

$$\frac{\text{\$ Annual demand charge, chillers}}{} = 2150 \text{ tons} \times \frac{.6 \text{ kW}}{\text{ton}} \times \frac{\$103}{\text{kW}}$$

where I have expressed the demand charge on an annual basis.

Dividing by the annual ton-hours produced by the Plant yields the Average Demand Charge/Ton-Hour.

$$\frac{\text{\$ Annual demand charge, chillers}}{\text{annual ton hours}} = \frac{2150 \text{ tons} \times \frac{.6 \text{ kW}}{\text{ton}} \times \frac{\$103}{\text{kW}}}{6{,}143{,}000 \text{ ton hours}}$$

Simplifying,

$$= \frac{1}{2837 \text{ hrs}} \times \frac{.6 \text{ kW}}{\text{ton}} \times \frac{\$103}{\text{kW}}$$

Note the ratio of peak tons to annual ton-hours—it's the "variable" in this equation.

Completing the calculation, we get 2.2 cents/ton-hour.

$$\frac{\text{\$ kW Demand}}{\text{ton-hour}} = 2.2¢/\text{ton-hour}$$

Returning to our equation for the cost/ton-hour to operate an electric centrifugal chiller at Grady Memorial Hospital and substituting for demand cost:

$$\frac{.6 \text{ kW}}{\text{ton}} \times \frac{2.5¢}{\text{kWh}} = \frac{1.5¢}{\text{ton-hour}} + \text{DEMAND} (=2.2¢/\text{ton-hour})$$

$$= \frac{3.7¢}{\text{ton-hour}}$$

Comparing 3.7 cents with the cost to operate any type of absorption chiller:

Single Stage....................$.10/ton-hour
Double Effect (Hitachi)...........$.059/ton-hour
Direct Fired....................$0.53/ton-hour

indicates a large cost/ton-hour advantage for the electric centrifugal machine, i.e., *the absorbers lose* at Grady Memorial Hospital.

Let's look at another example:

Concourse Office Building. Landmark's new Concourse Building is a 296,000-square-foot office building nearing completion of construction in Atlanta. It will operate on a typical 8-5, 5½-day-a-week business schedule. Computer simulation of the building's load profile and mechanical systems indicates that the building's chillers will run 3,075 hours during a typical year and produce 710 tons of cooling at peak load and 790,083 ton-hours altogether for the year. Thus,

$$\text{\$ Annual demand} = 710 \text{ tons} \times \frac{.6 \text{ kW}}{\text{ton}} \times \frac{\$86}{\text{kW}}$$

Calculating the average demand charge per ton-hour for the building:

$$\frac{\text{Annual demand charge chillers}}{\text{Annual ton-hours}} = \frac{710 \text{ tons} \times \frac{.6 \text{ kW}}{\text{ton}} \times \frac{\$86}{\text{kW}}}{790,083 \text{ ton-hours}}$$

$$= \frac{1}{1112 \text{ hrs}} \times \frac{.6 \text{ kW}}{\text{ton}} \times \frac{\$86}{\text{kW}}$$

Thus:

$$\frac{\text{\$ kW Demand}}{\text{ton-hour}} = 4.7 ¢/\text{ton-hour}$$

Note the following differences between this calculation and that for Grady Memorial Hospital.

(1) Due to different position on the Georgia Power PL-6 Rate Schedule:
  (a) Demand Charge is $86/kW-year[4] rather than $103/kW-year
  (b) kWh consumption charge is $.0273/kWh

(2) The ratio of peak tons to annual ton-hours produced is a much larger number—1/1112 versus 1/2857.

Can Absorption Chillers Beat Electric Centrifugal Chillers?

The average demand charge for Concourse of 4.7 cents/ton-hour is *much larger than that at Grady Hospital* (even though Grady pays more per kW of billing demand). In fact, the kW-demand portion of the electric cost is almost 75% of the total cost of producing a ton-hour of cooling.

The total cost/ton-hour to operate a .6 kW-ton electric centrifugal chiller at Concourse is:

$$\frac{.6 \text{ kW}}{\text{ton}} \times \frac{2.7\text{¢}}{\text{kWh}} = \frac{1.6\text{¢}}{\text{ton-hour}} + \text{Demand} (=4.7\text{¢/ton-hour})$$

$$= \frac{6.3\text{¢}}{\text{ton-hour}}$$

Comparing with the absorber choices—both the direct fired absorber and the Hitachi double effect steam absorber "appear to be" cheaper to operate.

So, if we are to simply answer the question posed at the beginning of this chapter, we have a dilemma. In some buildings, electric centrifugal chillers are clearly cheaper to operate; in other buildings, absorbers may be cheaper. Let's see if we can figure out why, and hopefully, generalize the results.

GENERALIZING THE RESULTS

In calculating the Average Demand Charge per ton-hour, the "variable" which caused the results to turn out so differently was the ratio of peak tons to annual ton-hours for each chiller plant.

$$\text{GRADY:} \quad \frac{\text{peak tons}}{\text{annual ton-hours}} = \frac{1}{2,875}$$

$$\text{CONCOURSE:} \quad \frac{\text{peak tons}}{\text{annual ton-hours}} = \frac{1}{1,112}$$

These ratios are, actually, the inverse of a more well-known parameter which is catalogued in a number of HVAC reference books, namely, the Equivalent Full Load Cooling Hours (EFLCHs). The EFLCHs are equal to annual ton-hours divided by peak load tons. You might think of it as: the number of hours a chiller plant would run at full-out peak load in order to produce the same number of ton-hours in a year as it actually produces at various part-loads.

$$\text{EFLCHs} = \frac{\text{Annual hours cooling}}{\text{peak tons}}$$

The formula we have used for calculating average kW Demand Charge per ton-hour can be simplified to the equation below where Demand Charge *as a function of EFLCHs*.

$$\frac{\$ \text{ Demand Charge}}{\text{ton-hour}} = \frac{1}{\text{EFLCHs}} \times \frac{.6 \text{ kW}}{\text{tons}} \times \frac{\$86}{\text{kW}}$$

Note the demand charge at $86/year has been fixed because it is typical of the type building which will have the most favorable load profile for absorption chillers, i.e., large commercial office buildings.

This equation is plotted on the figure below:

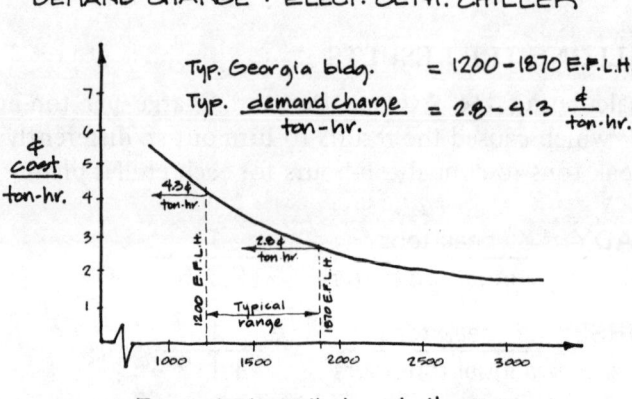

Figure 17-1. Equivalent full load hours.

Demand cost/ton-hour is on the vertical axis, while EFLCHs are laid out on the horizontal axis. We see that the building chilled water plants with a low number of Equivalent Full Load ton-hours will have a large demand charge per ton-hour while buildings which operate longer hours and hence have a larger number of EFLCHs will have a smaller demand charge/ton-hour. If we know a building's EFLCHs, then, we can estimate its demand charge per ton-hour.

Figures for EFLCHs for a number of U.S. Cities are listed in the ARHRAE Fundamentals chapter on Estimating Building Consumption, Table 2, Chapter 29. Another table published by the Johnson Control Company lists more cities in Georgia. Both references estimate the Equivalent "Rated" Full Load Cooling Hours for buildings in Georgia to be within the range from 1,200 hours to 1,870 hours.

For the range of EFLCHs applicable in Georgia (1,200-1,870), we see, by referring to Figure 17-1, that the demand charge will vary from a "typical" low of 2.8 cents/ton-hour to a "typical" high of 4.3 cents/ton-hour. In Figure 17-2 below, we include the kWh energy cost of 1.6 cents/ton-hour for the electric centrifugal chiller by moving our graph vertically 1.6 cents, then our y-axis represents the total cost per ton-hour to operate a .6 kW/ton electric centrifugal chiller for any EFLCHs. We see that the total cost/ton-hour to operate an electric chiller in a typical large commercial office building in Georgia is *4.4 cents to 5.9 cents/ton-hour.*

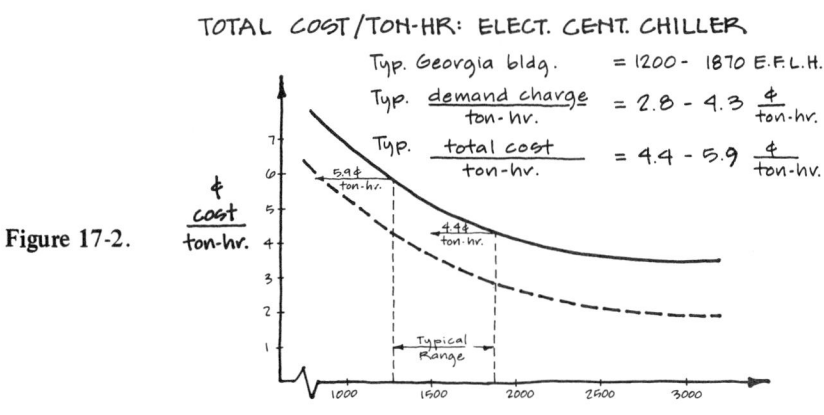

Figure 17-2.

The cost/ton-hour to operate a direct fired absorber is added in Figure 17-3 for comparison. The direct fired absorber, at 5.3 cents/ton-hour, is less expensive to run than an electric centrifugal chiller when EFLCHs are less than about 1,395 hours. The TRANE and Hitachi double effect steam absorption chillers, at 6.7 and 5.9 cents/ton-hour, exceed the cost of running an electric centrifugal chiller in most Georgia buildings.

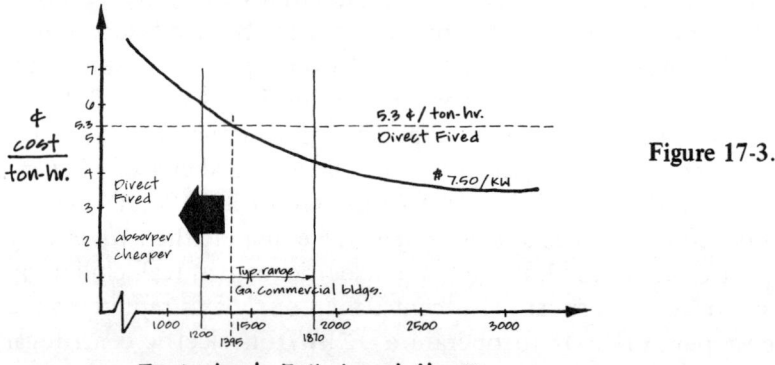

Figure 17-3.

So, we might conclude that: for some commercial buildings in Georgia, specifically those which operate for a limited number of hours per year—say, with a regular 8-to-5 office schedule—a direct fired absorption chiller may be cheaper to operate than a high efficiency electric centrifugal chiller.

This conclusion, however, would be hasty. We need to examine one more set of additional factors: What about parasitic and peripheral equipment run cost?

PARASITIC AND PERIPHERAL LOADS

Parasitic Loads

Parasitic loads consist of equipment items which are an integral part of the water chilling assembly and are necessary to its operation but are *not* included in its energy consumption or efficiency figures.

Can Absorption Chillers Beat Electric Centrifugal Chillers?

For a 700-ton chiller (the minimum size that would serve the 710-ton peak load in the Concourse Office Building), the following parasitic losses are estimated from manufacturer's catalogue data.

Parasitic Loads

	C-H-I-L-L-E-R T-Y-P-E		
	Elec. Cent.	Double Effect	Direct Fired
Parasitic Losses	1 kW	18 kW	23 kW

Where
1. The electric centrifugal chiller requires a ¾ HP oil pump.
2. The double effect and direct fired absorption chillers both require electric solution pumps to pump refrigerant and solvent from the absorber compartment to the concentrator.

Peripheral Loads

Peripheral loads consist of ancillary equipment which work in concert with the chiller. Generally, this equipment consists of chilled water pumps, condenser water pumps, and cooling tower fans.

The condenser pump and cooling tower fans are sized according to the GPM flow required by the condenser of each specific chiller type. The GPM, in turn, is determined by the amount of heat that must be removed from the building, plus, the heat required to generate the cooling effect. Since absorbers use heat energy to generate cooling, this heat must be rejected from the building in which the chiller sits. Absorbers, thus, use larger peripheral devices which consume more energy than those which would serve an electric centrifugal chiller, which must only reject electric motor heat. The ratio of absorber heat rejection to electric centrifugal heat rejection is calculated below on a per-ton basis.

HEAT REJECTION / TON

Ratio to 14,000

		Ratio to 14,000
$.6 \frac{KN}{ton}$ Elect. Cent.	= 12,000 Btu + .6(3413) Btu = 14,000	1.0
Double Effect	= 12,000 Btu + 10,600 Btu = 22,600	1.6
Direct Fired	= 12,000 Btu + .8(13,733) Btu = 23,000	1.6

Figure 17-4.

Parasitic and peripheral kW electric loads are tallied below. The first row of figures are the parasitic loads previously described. Calculation of the peripheral loads in the following two rows was accomplished as follows:

For the electric centrifugal chiller, the condenser water pump power draw of 28 kW reflects a 34 BHP pump sized for 3 GPM/ton of cooling and 50 feet of hydrostatic head. The cooling tower for the electric centrifugal chiller is sized in a likewise fashion. The peripheral load values for the absorbers are derived via the heat rejection ratios calculated above.

Peripheral/Parasitic Loads

C-H-I-L-L-E-R T-Y-P-E

	Elec. Cent.	Double Effect	Direct Fired
Parasitic Losses	1 kW	18 kW	23 kW
Cond. Pump	28 kW	33 kW	45 kW
Cooling Tower	29 kW	47 kW	47 kW
	58 kW	109 kW	114 kW

Some further discussion is in order here before continuing. I am assuming that the absorber's condenser water temperature rise will be 10 degrees, i.e., 85 to 95 degrees Fahrenheit, as would be typical for an electric centrifugal machine. Due to the greater amount of heat rejection, however, it would be advantageous to allow a larger temperature rise on the absorber's condenser, thereby decreasing the flow through the condenser and cooling tower. In fact, Hitachi does offer an optional condenser which permits a 15 degree temperature rise.

In this analysis, however, I am going to stick with a 10 degree rise for all machine types, because:

(1) In Hitachi's literature that's what they designated as "rated" conditions.

(2) An absorption chiller's performance does fall off with an increase in leaving conderser water, just as an electric centrifugal's would, and

(3) To permit the 15 degree rise, a special added option at a cost of about $10/ton is required if there is to be no degradation in performance.

Can Absorption Chillers Beat Electric Centrifugal Chillers?

The totals of peripheral and parasitic losses for a 700-ton chiller are 58 kW, 109 kW, and 114 kW. To express these totals in terms of cost/ton-hour, the annual electricity charge for these peripheral and parasitic losses is calculated and then divided by annual ton-hours using the formula:

$$\frac{\text{COST}}{\text{ton-hr}} = \frac{\text{kW} \times [(\text{hrs operation} \times \$.027/\text{kWh}) + 86\$/\text{kW}]}{\text{Annual ton-hours}}$$

Since hours of operation and annual ton-hours are building specific, the values for the Concourse Building, which we analyzed earlier, were substituted.

$$\frac{\text{COST}}{\text{ton-hr}} = \frac{\text{kW} \times [(\text{hrs operation} \times \$.027/\text{kWh}) + 86\$/\text{kW}]}{790,083 \text{ Annual ton-hours}}$$

The Concourse Building was chosen because its EHLCHs were calculated to be only 1113 . . . well below the minimum typical for a commercial building in Georgia. Thus, it appears to have a very favorable load profile for the use of direct fired absorption refrigeration.

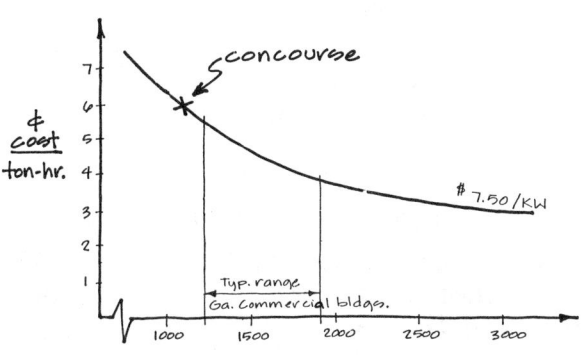

Figure 17-5.

Results and Discussion

The results of the calculation of parasitic and peripheral cost/ton-hour for the Concourse Building are shown on the following chart along with the costs we've previously calculated for the three chiller types.

RECAP - Cost/Ton-Hour

C-H-I-L-L-E-R T-Y-P-E-S

	Elec. Cent.	Double Effect	Direct Fired
Chiller Energy	1.6	5.9	5.3
Chiller Demand	4.6	—	—
Parasitic & Periph.	1.2	2.3	2.4
Total Cents = ton-hour	**7.5**	**8.2**	**7.7**

The final tally above shows that an electric centrifugal chiller, at .6 kW/ton, is, after all, cheaper to operate in the Concourse Office Building. The large parasitic and peripheral loads required by the absorption chillers caused their total run cost to overtake that of the electric centrifugal chiller.

While the run cost of the direct fired absorber is within 3% of that of the electric centrifugal chiller, it is not close to being competitive with it. The reason is first cost. The chart below lists the first costs of the chillers we are analyzing. Absorption chillers cost $110 more per ton than high efficiency electric centrifugal chillers.

1984 BID PRICE
based on 1,000-ton chillers

High Efficiency Electric Centrifugal	$150/ton
Double Effect Steam Absorber	$267/ton
Direct Fired Absorber	$264/ton

Thus, an owner purchasing a 700-ton absorption chiller in lieu of an electric centrifugal chiller would pay a premium of $77,000 before contractor mark-up and sales tax. Obviously, an absorption chiller must be significantly cheaper to operate if it is to pay back its additional capital cost.

Finally, it is important to keep in mind that this analysis was performed for a set of circumstances that was the *most advantageous* to an absorption chiller, This is because my purpose was to determine if an absorption chiller could, under any circumstances, beat the run cost of a high efficiency electric centrifugal chiller, assuming both chillers were base loaded.[6] Restating those circumstances which favored the absorber.

(1) Chillers were sized *exactly* to the cooling load. This minimized the size of peripheral and parasitic loads which, for an absorption chiller, require substantially more power than those of an electric centrifugal chiller. Oversizing the chillers, as would be typical to some extent in most buildings, would further penalize the absorption chillers in relation to the electric centrifugal chiller.

(2) Analysis was performed for the Concourse Office Building, whose Equivalent Full Load Cooling Hours are very low according to our simulation data, and lower than typical buildings in Georgia. This type of load profile accentuates electric peak demand charges while minimizing run time and ton-hours—ideal circumstances for an absorption chiller competing with an electric chiller.

CONCLUSION

Absorption chillers, operated as base loaded machines, are more expensive to operate than high efficiency electric centrifugal chillers in typical buildings in Georgia, under the current Georgia Power PL-6 Rate Schedule.

Postscript: The results of the previous analysis are sensitive to the relative costs of electricity and natural gas. If the price of electricity were to rise faster than that of natural gas, then direct fired absorption chillers would become more competitive in Georgia. This is expected to be the case when nuclear Plant Vogel begins generating commercial electrical power in 1987. Projections as of Spring

1985 predicted a 35% increase in the cost of electricity in 1987 due to the Plant. If, during the same period, natural gas rises 10% in price, as it is projected to do, the relative costs per ton-hour for the chiller alternatives will adjust as follows in the case of the Concourse Office Building.

	Elec. Cent.	Double Effect	Direct Fired
total run cost	10.1¢	9.6¢	9.1¢

At the run cost advantage of 1¢ per ton-hour, over 790,083 ton-hours, for the direct fired absorption chiller over the electric centrifugal chiller, the simple payback on the additional cost of the absorption chiller would be just under 10 years for the Concourse Office Building.

FOOTNOTES

[1] 40 cents/therm − 36.3 PGA + 3.5 cents/therm summer gas.

[2] 2.3 cents/kW is an average between 2.73, 2.35, 2.26; exact average is 2.45 cents/kWh without tax included.

[3] 8.97 x 11.45 months = $102.71/year x 1.05 (tax) = $107.85
400 < hours use < 600

[4] 200 < hours use < 400; $\frac{\$7.47}{kW} \times 11.45 \text{ months} = \frac{\$85.53}{kW/yr}$

[5] Note that the figures catalogued in both sources are based on rated design tons of a building's chiller plant rather than the actual building peak load (which is the basis of my definition). The "so what" of this difference is that the EFLCHs we are interested in will be 10-to-15% higher than the catalogued values. This tends to further bias this analysis in favor of absorbers.

[6] This analysis did not consider and does not pertain to mixed chilled water plants in which electric and absorption chillers may be sequenced to avoid electric peak demand charges while taking advantage of the electric chiller's low run cost during off-peak periods.

18

Humidification Steam Vs. Water

B.N. Gidwani

INTRODUCTION

Currently the HVAC systems which require winter humidification at Goddard Space Flight Center (GSFC) utilize an economizer cycle with steam as the source for humidification. Due to the continuously increasing cost of producing steam, a feasibility study was done at GSFC to evaluate alternate methods of humidification.

The result of this study is that the most economical system for humidification is atomization with deionization for water treatment.

DISCUSSION

The present heating, ventilation and air conditioning (HVAC) systems which require winter humidification operate on an economizer cycle using steam as the source of humidification. Due to the continuously increasing cost to produce steam, WESTON was asked to conduct a feasibility study to evaluate alternative methods of humidification at Goddard.

Two alternative methods were evaluated:

- Water-spray humidification with reverse osmosis water treatment using an economizer cycle.
- Water-spray humidification with deionization water treatment using the economizer cycle.

The energy, capital, operating and maintenance costs of the two alternatives were compared with:
- Steam humidification with the economizer cycle.
- Steam humidification with minimum outside air.

BASIS OF CALCULATIONS

In conducting the study, a few parameters were established:

1. The study is based on maintaining inside conditions of 72°F ± 2°F DB and 50% ± 10% relative humidity.

2. Building No. 25, with four HVAC systems and a total air flow of 120,000 cfm, was selected as being representative of all other buildings at the center.

3. One 30,000 cfm system within Building No. 25 was selected as being typical and is the basis for the following calculation: Humidification loads at maximum, average and lowest outdoor relative humidity with minimum outdoor air, and with the economizer cycle.

4. The weather data required for the calculations were determined from U.S. Department of Commerce weather data for Washington, D.C., for the year 1979. The data were compiled for 5°F temperature bins from 5°F to 72°F and for 3-hour intervals.

COST DATA

Utility costs are indicated below in terms of current fiscal year (FY) 81 costs (at the time of the study) and FY84 costs. To obtain the FY84 costs, the current costs, as provided by Goddard, were escalated at rates of 20% per year for fuel oil and sewerage, and 10% for water. The percentages for fuel oil, natural gas and electricity are based on NASA past experience, U.S. Department of Energy cost data, Energy Users News data, etc. Percentages for water and sewerage escalation are based on Goddard Space Flight Center cost data.

SYSTEM EVALUATION

Steam is the source of humidification at Goddard. The existing equipment including pneumatic controls are in good condition. No modifications can be made that would increase efficiency and thereby lower operating costs for steam humidification. Basically, there are only two alternatives to steam humidification:
- Atomization, which is commonly referred to as "water-spray."
- Evaporation.

Since evaporative-type humidification requires the addition of heat, it is inherently not as energy efficient as atomizing. Therefore atomization was the only alternative considered in this study.

WATER-SPRAY ATOMIZATION

Water-spray methods, whether injected directly or as a mist, are adiabatic processes. This is, no heat is transferred to or from the working media. Therefore, for the water-spray to be absorbed by the air stream, it must first be converted into a vapor by the addition of heat. Since the adiabatic process permits no heat to be added to the system, the water absorbs heat from the air stream in order to accomplish its vaporization. This sensible cooling of the air stream by the water-spray results in a lower leaving dry bulb temperature of the moist air mixture. This phenomena is illustrated by Figure 18-1.

This significant depression permits using outdoor air at higher temperature for cooling when the humidity is low. This is automatically done using an enthalpy controller. Hence there are significant additional savings during cooling season from water atomization.

There are two ways to atomize:

1. Direct Water Atomization—The water is supplied at high pressure to nozzles which atomize the water into a fine fog. The capacity of the nozzles vary with the pressure. Since it is difficult to modulate water pressure, the atomization capacity is controlled by sequencing the bank of nozzles depending on the humidification load.

2. Air Atomization—Water can also be atomized by using compressed air. Atomizing capacity is again controlled by sequencing the bank of nozzles depending on the humidification load. Even though the cost of air atomization is high due to use of compressed air, it is easier to control. The air atomizing nozzles have long been in use and are found to be extremely reliable.

Water-spray for humidification has the disadvantage of mineral "fallout" that results in a fine white powder that may be deposited on equipment in the conditioned space. This is especially critical in computer room applications. To prevent this, the water must be treated to remove minerals with two processes available: Reverse Osmosis or Deionization.

Reverse Osmosis

Reverse Osmosis is a process in which water under pressure (on the order of 60 to 600 psi) is forced through a semipermeable membrane. Newly developed membranes have controlled porosity which will reject most dissolved minerals, particulate matter and organics while allowing water to permeate through membrane. A typical system will produce 75% product water and 25% concentrated water solution.

Deionization

At Goddard Space Flight Center the hardness of water supply is relatively low favoring the use of deionization which is lower in capital, operating and maintenance costs. However, the deionized water is more aggressive than the effluent of reverse osmosis, and the system piping, spray nozzles and other components must be compatible. Plastic piping with stainless steel nozzles have been commonly used. The water must also be filtered to avoid clogging the nozzles.

RESULTS AND CONCLUSIONS

Figure 18-2 shows the summary of operating costs for the 30,000 cfm systems at various conditions. As noted, the costs of steam humidification with the economizer cycle are relatively high.

These costs are reduced considerably if the relative humidity requirements are dropped from 50% to 40%.

Figure 18-3 shows the same comparison, except that utility costs are escalated to FY84.

An evaluation of water-spray humidification systems for individual buildings and central systems for several buildings indicates that the central system would be lower in first cost as well as operating and maintenance costs. Figure 18-4 presents a cost comparison of steam and water-spray humidification for individual buildings and central-type systems.

The result of WESTON's study for NASA's Goddard Space Flight Center is that the most economical system is atomization with deionization for water treatment. Estimated FY84 energy savings with the atomizing system is 83.1×10^9 source Btu. This is about 4.3% of Goddard's 1973 base year energy consumption of 1970×10^9 source Btu (Tables 18-1 through 18-5).

For the entire Goddard Space Flight Center complex, the estimated FY84 savings compared to the present steam humidification system is $700,000 annually. The estimated FY84 installation cost is $390,000 with a simple payback of .57 years.

Allowing for estimated annual cost of $100,000 for Goddard system maintenance, spare parts, technical documentation, etc. increases the simple payback period to .64 years. If equipment costs were to escalate at 20% per year instead of the manufacturer's recommended 10% per year, the effect on the payback period would be negligible.

Therefore, WESTON recommended that NASA should replace the existing steam-humidification system with an air-atomizing-type system using deionization for water treatment. A pilot building was recommended to verify the results.

NASA FY81 & FY84 Utility Costs

- **Fuel Oil:** Current — $1.00 Per Gallon
  FY84 — $1.95 Per Gallon

- **Natural Gas:** Current — $0.40 Per Therm
  FY84 — $0.70 Per Therm

- **Steam:** Current — $6.16 Per Thousand Pounds
  FY84 — $11.16 Per Thousand Pounds

- **Electricity:** Current — $ .05 Per Kilowatt Hour
  FY84 — $ .086 Per Kilowatt Hour

- **Water:** Current — $1.20 Per Thousand Gallons
  FY84 — $1.60 Per Thousand Gallons

- **Sewerage:** Current — $2.00 Per Thousand Gallons
  FY84 — $3.90 Per Thousand Gallons

Figure 18-1. Water Spray Humidification
Adiabatic Process

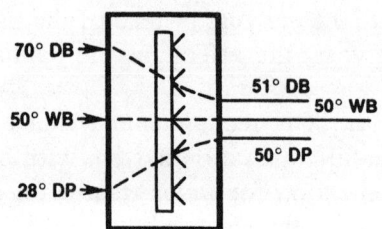

Total Heat Remains Constant
Sensible Temperature Drops
Moisture Addition 30 grains/pound

Humidification Steam Vs. Water

Figure 18-2. Summary of Operating Costs for 30,000 CFM System Operating at Various Conditions.

**Costs Are For FY81
For 72°F. & 50% RH**

System Operation	Annual Cooling Cost During Econo. Cycle Hrs.* #/Yrs.	Annual Cost of Humid.	Total Annual Cost	Annual Savings vs. Present System
No Outdoor Air	9,570	0	9,570	170
Min. O.A. w/Steam Humid.	9,110	455	9,565	175
Econo. Cycle w/Steam Humid.	910	8,830	9740	(BASE)
Econo. Cycle w/Water Spray & Rev. Osmosis	140	2,690	2,830	6,910
Econo. Cycle w/Water Spray & Deionization	140	580	720	9,020

*Hours When Humidification is Required

Figure 18-3. Summary of Operating Costs for 30,000 CFM System Opearting at Various Conditions

**Costs Are For FY84
For 72°F. & 50% RH**

System Operation	Annual Cooling Cost During Econo. Cycle Hrs.* #/Yrs.	Annual Cost of Humid.	Total Annual Cost	Annual Savings vs. Present System
No Outdoor Air	16,460	0	16,460	1,150
Min. O.A. w/Steam Humid.	15,670	825	16,495	1,050
Econo. Cycle w/Steam Humid.	1,565	16,000	17,565	(BASE)
Econo. Cycle w/Water Spray & Rev. Osmosis	240	3,790	4,030	13,535
Econo. Cycle w/Water Spray & Deionization	240	865	1,105	19,460

*Hours When Humidification is Required

Figure 18-4. Comparison of Steam vs. Water-Spray Humidification

Bldg. No. or Bldg. Complex	Annual Stm. Load for Humid. Lbs x 10³	FY84 Steam Cost	FY84 Installed Cost for Water Spray w/D.I.	FY84 Operating Cost for Water Spray w/D.I.	Annual Savings Water Spray vs. Stm. Humid.	Simple Payback Yrs.
3	4,780	53,345	35,800	2,880	50,465	.71
7	11,376	126,960	83,400	6,860	120,100	.69
10	2,868	32,000	26,850	1,730	30,270	.88
13	1,816	20,270	18,500	1,100	19,170	.96
14	11,520	128,560	83,400	6,950	121,610	.69
15	1,147	12,400	18,500	690	11,710	1.60
22	3,226	36,000	26,850	1,945	34,055	.79
23	13,000	145,100	83,400	7,840	137,260	.60
25	5,736	64,000	35,800	3,460	60,540	.59
28	11,050	123,330	83,400	6,670	116,660	.71
3/13/14	18,116	202,375	90,000	10,930	191,445	.47
7/10/15	15,391	171,360	90,000	9,285	162,075	.56
22/23	16,226	181,100	90,000	9,800	171,300	.53

Note: Operating Cost for Water Spray Does Not Include Allowance of $100,000.00 for Annual GSFC Maintenance, Spare Parts, Technical Documentation.

Table 18-1. Hourly Temperatures and CFM Data for 30,000 CFM on Economizer Cycle of Operation.

(1) Outdoor Air Temp. Bin	(2) Av. Bin Temp.	(3) Fraction of O.A. Req. on Econo. Cycle	(4) O.A. CFM
70-72	71	1.0	30,000
65-69	67	1.0	30,000
60-64	62	1.0	30,000
55-59	57	1.0	30,000
50-54	52	.850	25,500
45-49	47	.680	20,400
40-44	42	.567	17,010
35-39	37	.486	14,580
30-34	32	.425	12,750
25-29	27	.378	11,340
20-24	22	.340	10,200
15-19	17	.309	9,270
10-14	12	.283	8,490
5-9	7	.262	7,860

Humidification Steam Vs. Water 341

Table 18-2. Humidification Requirements for 30,000 CFM System at Avg. Outdoor RH and Economizer Cycle of Operation

(1) O.A. Temp. Bin	(5) Av. O.A. % RH	(6) Hum. Ratio @ Satur. W_s lb. H_2O / lb. Air	(6A) Sp. Vol. Satur. Air Vs. ft^3/lb.	(7) Gr./ft^3 @ Satur. $\frac{(6)}{(6A)} \times 7000$	(8) Gr./ft^3 @ Av. RH (7) x (5)	(9) Inside Air Gr./ft^3 @ 72-50%	(10) Water Req. Gr./ft O.A. (9)-(8)	(11) Water Req. lb./hr. (10) x (3) x 257.14	(12) Hr./Yr. In Temp Bin	(13) Water Req. lb./Yr. (11) x (12)
70-72	43.1	.0163	13.72	8.36	3.60	4.32	0.72	185	66	12,200
65-69	45.0	.0142	13.57	7.34	3.30	4.32	1.02	262	189	49,500
60-64	51.9	.0119	13.39	6.22	3.23	4.32	1.09	280	309	86,500
55-59	60.6	.0099	13.23	5.26	3.19	4.32	1.13	290	504	146,200
50-54	66.3	.0083	13.06	4.42	2.93	4.32	1.39	304	756	230,000
45-49	68.8	.0068	12.91	3.71	2.48	4.32	1.84	322	684	220,000
40-44	67.5	.0056	12.78	3.09	2.09	4.32	2.23	325	594	193,000
35-39	66.8	.0046	12.61	2.57	1.72	4.32	2.60	325	570	185,000
30-34	54.9	.0038	12.46	2.13	1.17	4.32	3.15	344	417	143,000
25-29	56.0	.0030	12.32	1.71	0.96	4.32	3.36	327	198	65,000
20-24	58.0	.0024	12.18	1.36	0.79	4.32	3.53	309	165	51,000
15-19	48.7	.0019	12.04	1.08	0.53	4.32	3.79	301	96	29,000
10-14	55.2	.0015	11.91	0.856	0.47	4.32	3.85	280	69	19,000
5-9	42.3	.0011	11.78	0.672	0.28	4.32	4.04	272	18	5,000
									4629	1,434,000

Table 18-3. Cooling Savings for 30,000 CFM on Economizer Cycle of Operation

(1) Outdoor Air Temp Bin	(27) Cooling Saving Tons	(28) KWH/Yr (27) x 1.0 x (12)	(29) $/Yr (28) x .05
70-72	2.7	200	10
65-69	13.5	2,300	120
60-64	27.0	7,500	380
55-59	40.5	18,400	920
50-54	45.9	31,200	1,560
45-49	45.9	28,300	1,420
40-44	45.9	24,500	1,220
35-39	45.9	23,500	1,180
30-34	45.9	17,200	860
25-29	45.9	8,200	410
20-24	45.9	6,800	340
15-19	45.9	4,000	200
10-14	45.9	2,800	140
5-9	45.9	700	40
		175,800	8,800

Table 18-4. Temperature Reduction with Water Spray – 30,000 CFM System – Economizer Cycle – Av. Outdoor R.H.

(1) Outdoor Temp Bin	(34) Mixed Air Temp. °F (2) - .0329 x (11)*	(35) Temp. Depress (2) - (34)	(36) Fraction Outside Air	(37) O.A. CFM (36) x 30,000	(38) Water Required Lbs/Hr (10) x (36) x 257.14	(39) Water Required Lbs/Yr (38) x (12)
70-72	64.9	6.1	1.00	30,000	185	12,200
65-69	58.4	8.6	1.00	30,000	262	49,500
60-64	63.1	8.1	.889	26,670	249	76,900
55-59	61.7	6.7	.688	20,640	200	100,800
50-54	61.3	6.3	.534	16,020	191	144,400
45-49	61.5	6.5	.418	12,550	198	135,400
40-44	61.6	6.6	.348	10,450	200	118,800
35-39	61.6	6.6	.297	8,910	199	113,400
30-34	61.8	6.8	.254	7,620	206	85,900
25-29	61.6	6.6	.232	6,960	200	39,600
20-24	61.3	6.3	.214	6,420	194	32,000
15-19	61.2	6.2	.196	5,870	191	18,300
10-14	61.0	6.0	.183	5,500	181	12,500
5-9	60.9	5.9	.171	5,140	178	3,200
						942,900

*Only for Temp. Above 55°

Table 18-5. Additional Cooling Savings with Water Spray Due to Temp. Depression – 30,000 CFM – Economizer Cycle

(1) Outdoor Temp Bin	(40) Savings Tons (35) x 2.7	(41) Savings KWH/Yr (40) x .9 x (12)	(42) Savings $/Yr (41) x .05	(43) Total Savings $/Yr (42) + (29)
70-72	16.5	980	50	60
65-69	23.2	3950	200	320
60-64	18.9	5260	260	640
55-59	5.4	2450	120	1040
50-54	-	-	-	1560
45-49	-	-	-	1420
40-44	-	-	-	1220
35-39	-	-	-	1180
30-34	-	-	-	860
25-29	-	-	-	410
20-24	-	-	-	340
15-19	-	-	-	200
10-14	-	-	-	140
5-9	-	-	-	40

19

Enthalpy Vs. Dry Bulb Start Time Optimization

B.D. Mayfield

I. **What is Start Time Optimization.**

Start Time Optimization (STO) is defined as a program or software control to allow for the most economical start-up of a building. The objective is to consider inside conditions, outside conditions, and desired inside conditions at occupancy, using this information to bring the building up using the least amount of energy. The program will adjust the start times of various equipment based on the ability of the mechanical system and the thermal inertia of the building.

II. **How Most Start Time Optimization Programs Work.**

The description provided below is not intended to describe all Start Time Optimization Programs but is meant to be an overview and a means of understanding what takes place within the program.

A) **Dry Bulb:**
The Dry Bulb type optimization program works with various variables which are described below:

1) Occupancy Time — The actual time which the area being controlled will be occupied.

2) Rate Parameter — Represents the time required to change the temperature one degree for the occupied area.

3) Desired space temperature — that temperature which the operator desires the occupied area to be at when occupancy time occurs.

4) Present Area Temperature — Obtained from sensor or sensors.

5) Outside Temperature — Obtained from sensor or sensors.

6) Release Time — The time which the STO program is released to run.

The STO program will be given a release time which allows the program to start the mechanical equipment required to bring the area to desired conditions any time after passing the release time.

The software will, after release time is passed, check inside temperature, outside temperature vs. desired temperature and will multiply degrees per minute change required and then subtract this from occupancy time to come up with the time it needs to start the equipment. An example is shown below for a warm-up condition.

Occupancy Time: 8:00 A.M.
Rate Parameter: Inside = 4 minutes/degree
                   Outside = 3 minutes/degree
(Difference calculated from inside)

Desired Space Temperature: 68 degrees

Present Area Temperature: 55 degrees

Outside Temperature: 40 degrees

Step 1
(68 − 55) X 4 = 52
(Desired − Present) X minutes/degree = 52 minutes

Step 2
(55 − 40) X 3 = 45
(Inside − Outside) X minutes/degree = 45 minutes

Step 3
52 + 45 = 97
(Inside warm-up + Outside offset) = total time for early start

Step 4
8:00 A.M. − 97 minutes = 6:23 A.M.
Present Time − minutes for start up = Optimized Start Time

When 6:23 A.M. is reached the mechanical system which is required to warm up the area will be started.

At this point I should add that this type of calculation will be run every 15 minutes from the release time given. The software will also note the time desired temperature was reached and also the inside temperature at occupancy. It will then use this data to go back and self correct the program.

Normally the following parameters will allow for selection of different minutes/degree tables based on conditions.

Inside Vs. Desired Vs. Outside

	Inside	Desired	Outside
Table 1		<	<
2		<	>
3		=	>
4		=	<
5		>	<
6		>	<

B) **Enthalpy**

This program works in the same basic manner as the Dry Bulb with the difference being that Enthalpy values are used for inside and outside calculations. The building will be given a Btu/lb value for desired conditions rather than a dry bulb temperature.

III. Psychrometric Chart Comparison of Enthalpy Vs. Dry Bulb

A) **Heating Condition using free outdoor air heat:**
Conditions:
Inside: 65 F/50% RH
Desired: 68 F/50% RH
Outside: 80 F/80% RH

The psychrometric chart shows that under Enthalpy or Dry Bulb optimization, free heat can be used until outside RH reaches approximately 35%, mechanical heat will be required if strict Enthalpy control is used.

B) **Cooling Condition using free outdoor air cooling:**
Conditions:
Inside: 80 F/50% RH
Desired: 78 F/50% RH
Outside: 70 F/80% RH

The psychrometric chart shows that under Enthalpy or Dry Bulb optimization free cooling can be used until outside RH reaches approximately 70%, mechanical cooling will be required if strict Enthalpy control is used.

IV. Conclusion and Summary

One should pay close attention to climatic conditions in the area in which the optimization program is to be applied. If the Start Time Optimization Program is applied improperly, it will cost energy rather than save energy. You should also remember that the control system will be calibrated to dry bulb temperature and will control to dry bulb, not Enthalpy.

Enthalpy Vs. Dry Bulb Start Time Optimization 347

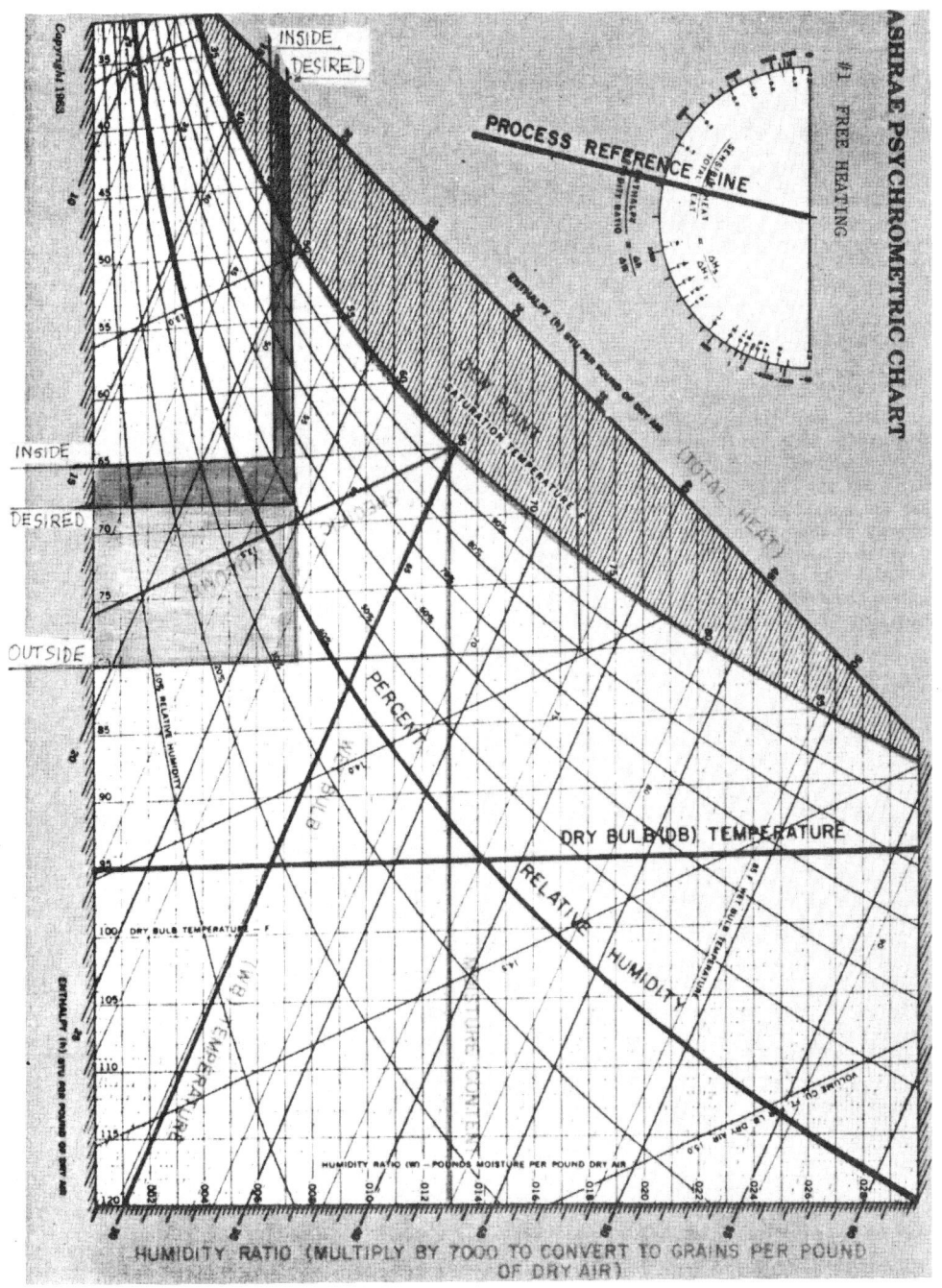

OPTIMIZING HVAC SYSTEMS

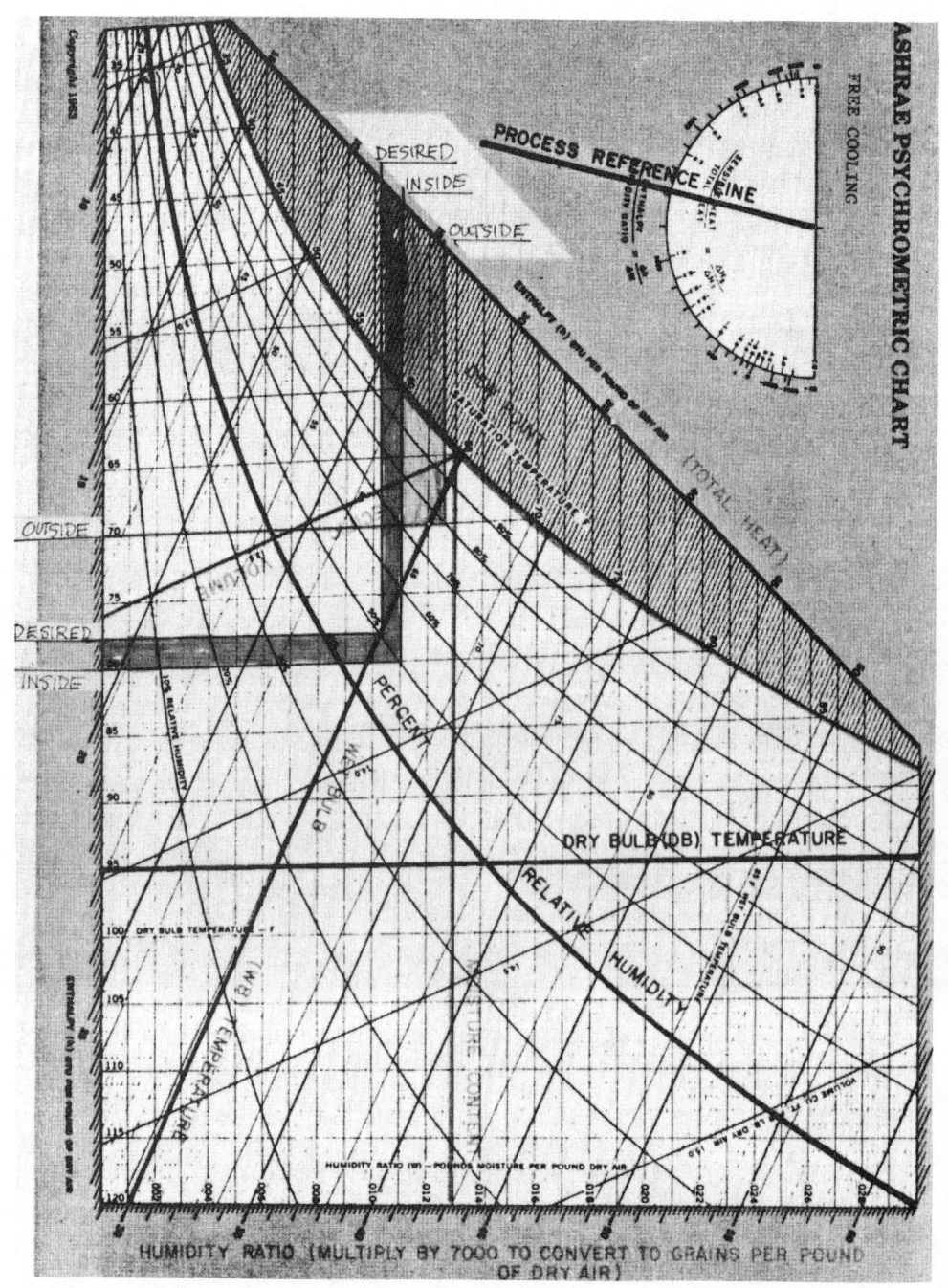

20

Technology Update for Desiccant-Based Air Conditioning Systems

P.J. Banks

INTRODUCTION

In recent years, increasing electrical costs, including rising demand penalities, have forced owners, design engineers, and manufacturers to search for alternative methods of air conditioning. This, combined with increased research being sponsored by the Gas Research Institute, the Solar Energy Research Institute and the Department of Energy has resulted in some interesting alternatives to conventional electrically driven, vapor-compression air conditioning systems. One of the most promising alternatives is the utilization of desiccant-based air conditioning systems.

DESICCANTS

Desiccants are absorbent or adsorbent materials which remove water or water vapor from another material. An absorbent desiccant removes moisture and in so doing, undergoes a chemical change in a manner analogous to dry milk being mixed with water. On the other hand, an adsorbent desiccant does not undergo a chemical change but accepts the water molecules through strong molecular attraction similar to a sponge soaking up water. Although most people are familiar only with the solid silica gel desiccant found in packaging for cameras and electronic products, desiccants are available in several different forms including both liquid and solid.

Desiccant systems have held a strong position in the industrial marketplace, being utilized in the pharmaceutical, chemical, food product, and many other manufacturing and storage industries. The desiccant-based system application has been primarily dictated by design requirements of very low specific humidity ratios (less than 45 gr/lb.). The primary advantage of a desiccant system is its ability to efficiently dry air to very low humidity levels. Attempting to meet these very low specific humidity levels with conventional, electrically driven vapor compression (V.C.) air conditioning equipment—the major competitor of desiccant systems—can prove quite expensive in both initial equipment and installed operating costs. Accomplishing this task with conventional V.C. refrigeration often entails the use of parallel sets of compressors. The frosting problems associated with low suction temperatures requires a parallel operation to allow one compressor to handle the load while the other defrosts. Furthermore, operating at low suction temperatures substantially decreases the coefficient of performance (COP) and efficiency of vapor compression systems. Attempts to improve overall system efficiency by using series compressors (one operating at a higher suction temperature) increases both the initial and maintenance costs of the system.

Desiccant Process

A typical desiccant dehumidification process begins by passing a moisture-laden air stream through a liquid or solid desiccant material which extracts the moisture from the air. In a process called regeneration, the desiccant material is then heated in order to drive the moisture out of the system. Once the desiccant has been regenerated, it may be reutilized to remove more moisture. In fact, most desiccants are capable of being regenerated thousands of times before requiring a chemical revitalization or replacement.

A desiccant which has been popular in industry for many years is the rotary wheel dehumidifier shown in simplified form in Figure 20-1. A wheel, composed of a fiberglass structure, is impregnated with a desiccant such as lithium chloride. The wheel rotates at a slow rate (7 to 10 rph) within a cylindrical casing and simultaneously dehumidifies as it undergoes regeneration. This is accomplished by utilizing flexible rubber seals to separate the drying and regenerating sectors and air streams in a counterflow arrangement. This psychometric process is compared to a conventional vapor compression dehumidification process in Figure 20-2. Since the moisture removal capacity is dependent upon the air regeneration temperature and flow rate, the capacity of the unit drops off as the regeneration temperature decreases.

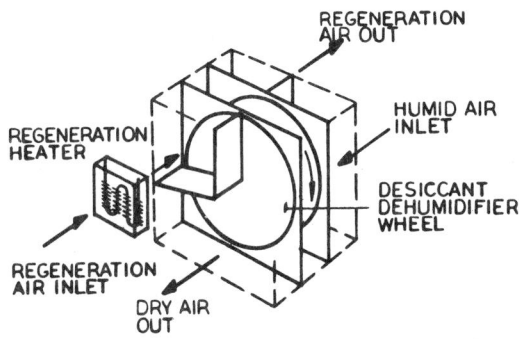

Figure 20-1. Rotary wheel dehumidifier.

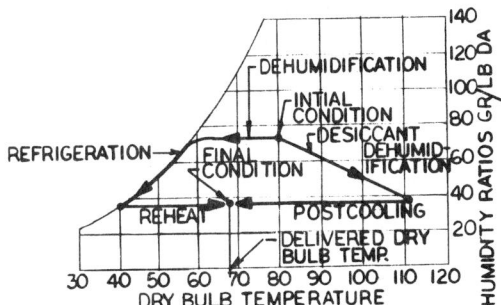

Figure 20-2. Psychometric chart showing vapor compression dehumidification and desiccant processes.

Although typical regeneration temperatures for industrial desiccant systems require high grade energy sources, desiccant-based systems are now being employed with lower quality energy from solar and heat recovery. Typical regeneration temperatures now being used (140°F-220°F) are within the boundaries of cogeneration engines, waste product boilers, refrigeration heat recovery systems, and solar energy. Low regeneration temperatures can be compensated for by increasing the wheel size. Desiccant manufacturers are currently working toward developing wheels which do not require high temperature regeneration cycles.

THE MARKETS

Although extremely successful and widely accepted in the industrial marketplace, desiccant-based systems have not played a major role in the residential or commercial markets. This is primarily due to low electrical rates and the ability of conventional air conditioning systems to efficiently handle the high sensible/total heat ratios typically found in these market sectors. This is, however, changing with the dynamic utility rate structures and the increased implementation of systems employing heat recovery and other low cost sources of energy. A brief look at actual and hypothetical situations in all three marketplaces (residential, commercial, and industrial) will illustrate some of the factors involved in the newfound success of desiccant-based air conditioning systems.

Residential

Until recently, the residential air conditioning industry has not felt the crunch of rising energy costs as did its counterpart heating market. Shortly after the energy crisis in the mid-seventies, the heating equipment manufacturers responded by improving heating system efficiencies and downsizing equipment to meet the decreasing system requirements of efficient, well-insulated solar homes. Recent stabilized oil and gas prices have decreased the demand for active solar systems. However, solar-based, desiccant air conditioning systems may revitalize the solar industry. American Solar King of Waco, Texas, is developing and marketing a desiccant-based system which uses solar heated water to regenerate a desiccant wheel, in conjunction with a heat recovery wheel and a direct evaporative cooler. A flow schematic of the system components is shown in Figure 20-3.

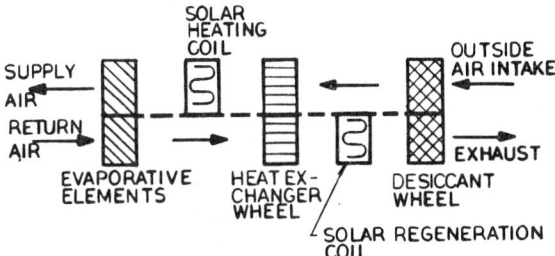

Figure 20-3. System schematic residential solar/desiccant air conditioning system.

The concept behind the system is to dry outside air to a very low specific humidity level (providing latent cooling) and then direct the dry hot air through an evaporative cooler in order to deliver the air at comfort levels. The system uses 100 percent outdoor air to provide five to ten air changes per hour and approximately three tons of central air conditioning. To maximize efficiency, the system employs a rotary heat exchanger which simultaneously post-cools the desiccant wheel and pre-heats the regeneration air. It is interesting to note that one sector of the residential marketplace currently targeted is the new, efficient "tight" homes in which indoor air pollution is being recognized as a health hazard. The system, which uses less electricity than most small appliances, is expected to provide annual savings of 60 to 80 percent over its conventional counterparts.

Commercial

There are a number of promising sectors in the commercial marketplace for desiccant systems. One application which has been researched quite thoroughly and is proving to be successful is the supermarket. Supermarkets provide an interesting example of an energy intensive environment for which engineering ingenuity, combined with the desiccant technology, is achieving considerable cost reductions for the owner.

In 1981, the median annual energy expense for a group of supermarkets surveyed by the Good Marketing Institute was $3.10/

square foot.[1] This relatively high operating cost is reasonable given the complexities of the supermarket environment. With its numerous refrigerated cases, a typical supermarket environment experiences both latent and sensible cooling effects. The amount of cooling realized is dependent upon the number, length, and type of case in the store. A credit is usually taken in store H.V.A.C. design for the sensible effect as well as any latent benefits realized although, in fact, the sensible and latent cooling derived from the refrigerated cases is not energy efficient for several reasons. Most significant is the poor C.O.P. of medium and low temperature refrigeration cases which operate at low suction temperatures ($-35°F + 35°F$) in comparison to conventional H.V.A.C. vapor compression machines ($40°F$-$50°F$). In fact, as the refrigerated cases indirectly provide moisture removal from the space, frost forms on the refrigerant coils (due to their low operating temperatures) and condensate fogs up glazed case doors. This frosting and condensation requires an inefficient refrigeration defrost cycle.

The basic concept behind utilizing desiccant-based H.V.A.C. systems monopolizes these inefficiencies by lowering the specific humidity of a store below the dewpoint of the medium case and reducing the defrost cycle time on the low temperature cases. A desiccant-based supermarket system flow schematic is shown in Figure 20-4. Several different supermarket chains nationwide utilize this particular system which is offered by Cargocaire Engineering Corporation of Amesbury, Massachusetts. The system takes advantage of heat reclaim from the food case refrigerant hot gas for winter heat (a typical practice in state-of-the-art conventional systems) and minimizes the costs associated with post cooling and regeneration by using a desuperheating coil to preheat regeneration air.

A system flow schematic comparison for a typical 30,000-square-foot supermarket located in Boston, Massachusetts, is shown in Figure 20-5, with an analysis of the annual operating costs of each system summarized in Table 20-1. It should be noted that the reduced CFM/Sq. Ft. used by the desiccant system is primarily because the entire latent load is now handled by the dehumidifier and that the air is dried so deeply that the quantity of CFM delivered is reduced to the minimum amount required to handle the sensible load.

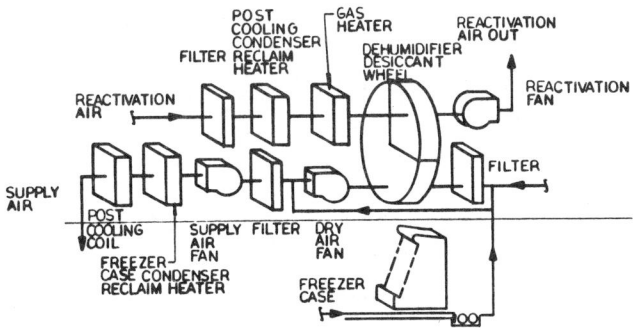

Figure 20-4. System schematic showing desiccant based air conditioning systems as applied in supermarkets.

Summary of Analysis Annual Operation of Conventional Air Conditioning Versus Cargocaire Superaire 1 System for Boston, Massachusetts, 30,000 ft^2 typical sypermarket, with electricity at \$0.064 kWh and gas \$0.50 therm and 75°F store temperature.

1. Case Energy Savings Versus Relative Humidity

% RH	\$ Case Credit
55	0 (base point)
50	732
45	1,763
40	2,929
35	4,257
30	5,722

2. A/C Energy Costs, Case Credits and Net A/C Energy Costs

System	scfm/ft^2	\$ RH	\$ Cost	\$ Case Credit	\$ Net A/C Cost
Conventional	0.7	55	10,060	0 =	\$10,060
Superaire I	0.5	40	7,498	2,929 =	4,569
Superaire I	0.5	30	8,802	5,722 =	3,080

Table 20-1.

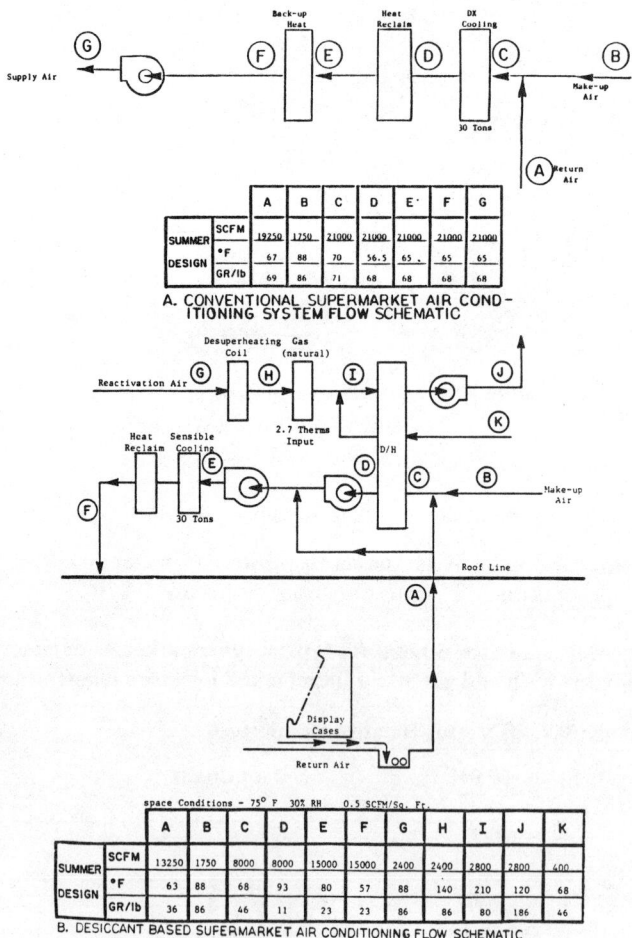

Figure 20-5. Flow schematics — conventional versus desiccant supermarket HVAC systems.

Currently, a typical equipment premium of 10-20 percent above conventional system costs for the desiccant system yields a 2- to 3-year payback in most comparisons. This payback is improved when credit is taken for reduced ductwork and refrigerated case size requirements. Another advantage, more difficult to quantify, is improved product visibility and a more comfortable store environment due to increased aisle air temperatures. However, designers should proceed with caution since payback results are dependent on the quantity of refrigerated cases, sensible loads (envelope, lighting, baking, and people) and latent loads (people and outdoor air).

Perhaps the most promising market for desiccant-based systems will be in commercial buildings. The Gas Research Institute is sponsoring research in this area and has backed the development of a prototype desiccant vapor compressor system similar to the supermarket system. An example of a simple desiccant-based air conditioning system which eliminates the requirements for the vapor compression system is shown in Figure 20-6. The viability of such a system relies on the climatic location of the building, the cost of utilities, the effective sensible heat to total heat ratio, and the availability of recoverable waste heat for regeneration.

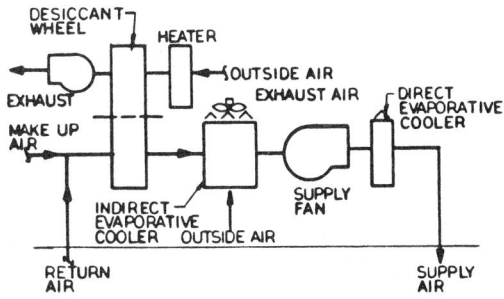

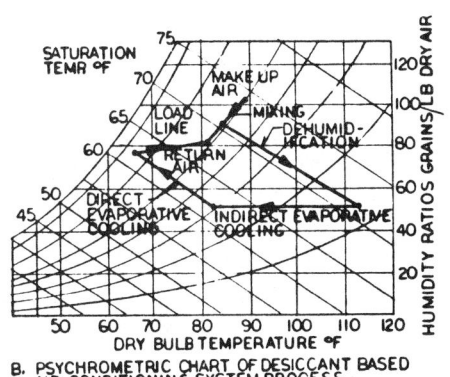

Figure 20-6. System flow schematic and psychrometric process for desiccant-based commercial air conditioning system.

When examined in a total system analysis, a desiccant system offers several unique initial cost and operating costs advantages as proven by Mecklar.[2] Mr. Mecklar has investigated and implemented several innovative designs involving desiccants. One of the most interesting designs by Mecklar was for a large (786,000-Sq.-Ft.) office building which had significant internal gains from computers. Design required sizing the refrigeration plant to handle this large internal load, anticipated to be spread throughout an unspecified (at the time of design) 40 percent of the building. A hybrid desiccant/chilled water system was investigated. A two-stage liquid desiccant dehumidifier is used to deeply dry the minimum required outdoor (ventilation) air. This dry air is ducted through the building and has dried to a point which satisfies the full latent load. This, combined with the flexibility of a modular piping/fan coil system, provided the lowest life cycle cost analysis. The system offered the following advantages:

- Reduced initial ductwork costs.
- Decreased floor-to-ceiling requirements.
- Reduced fan power requirements.
- Increased flexibility and reduced cost for interior space changes.
- Reduced refrigeration requirements.
- Increased refrigeration efficiency due to improved C.O.P. since low suction temperature requirements are not necessary for dehumidification.

Industrial

Although desiccants are already popular in the industrial market, new applications will appear as the commercially manufactured units become available (as opposed to expensive built-up industrial units) and as electrical prices continue to increase.

For example, in 1982, the author investigated the utilization of a desiccant-based air conditioning system for a virus-free laboratory which had high internal sensible and latent loads and required 100% outdoor air. The existing system was extremely energy intensive;

high internal gains and outdoor air load resulted in a low effective sensible heat to total heat ratio of 0.5. System schematics are shown with respective psychrometic diagrams shown in Figures 20-7 and 20-8. The comparison for this particular installation yielded the following results:

SYSTEM COMPARISON AT SUMMER DESIGN CONDITIONS

	Gas Heat Required BTUH	Gas Heat* $/Hr	Total Direct Expansion Cooling Required	Electrical Cost $/Hr	Initial Equipment Cost
System 1 (Conventional)	203,500	$1.42	53 Tons	$7.45	$83,000
System 2 (Desiccant)	483,800	$3.40	39 Tons	$5.50	$127,000

Assumptions:
$0.10/kWh average electricity costs.
$0.70/therm average natural gas costs.
Direct expansion cooling COP = 2.5

This preliminary evaluation assumed that the winter operation of both systems would be identical in energy costs (both utilizing outdoor air for cooling) and that system efficiencies would drop off equally at part load. The evaluation never went beyond this stage because of the high initial cost of the desiccant-based system and the equally high operating costs.

The desiccant-based system was considered to have limited advantages, such as certain bactericidal properties associated with the lithium chloride desiccant, the fact that the cooling coil operated dry (non-condensing), as well as the ability to reduce initial costs for installations where there was limited electrical power available. The high initial costs of the desiccant system were attributed to a built-up system which utilized an integrated industrial dehumidifier. Today, however, initial costs for integrated dehumidification systems are closing in on conventional systems as manufacturers adapt to the commercial marketplace. A lower initial equipment cost, augmented by today's increasing electrical energy costs (particularly rising demand charges), and the introduction of the ability to utilize the waste heat from a cogeneration plant for regeneration heat would shed a new light on this evaluation.

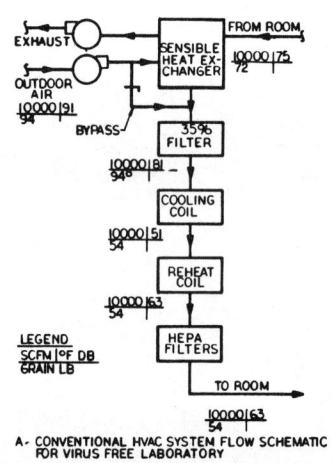

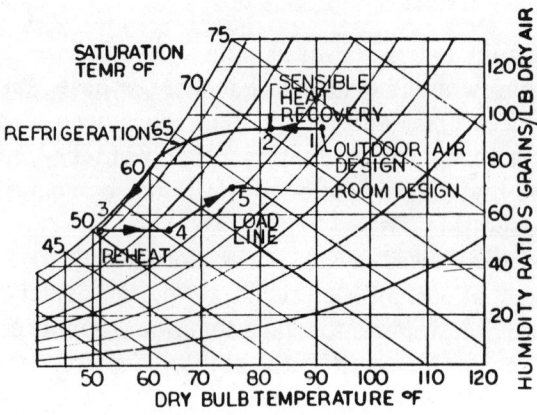

Figure 20-7. System flow schematic and psychrometric process for conventional air conditioning system — virus free laboratory application.

Technology Update for Desiccant-Based Air Conditioning Systems

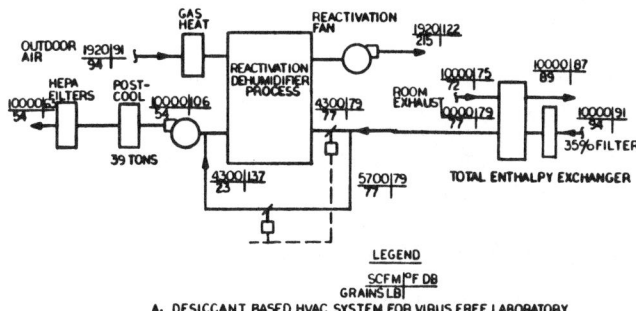

A. DESICCANT BASED HVAC SYSTEM FOR VIRUS FREE LABORATORY

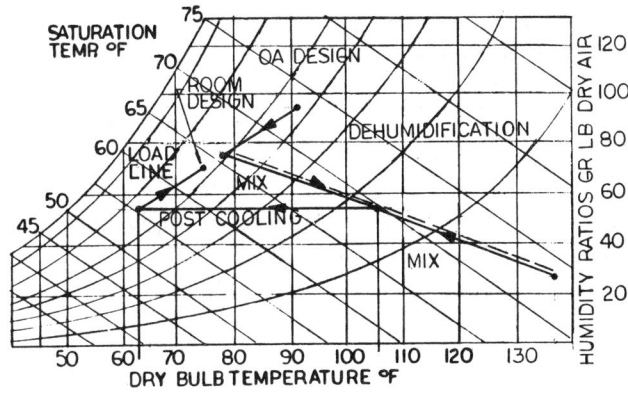

B PSYCHROMETRIC PROCESS

Figure 20-8. System flow schematic and psychrometric process for desiccant-based air conditioning system — virus free laboratory application.

CONCLUSION

The ability of desiccant-based HVAC systems to efficiently dehumidify to low humidity levels has been recognized for many years. Recent research, which has been fueled by the search for alternatives to high cost, electrically driven vapor compression machines, has produced several viable market sectors for new desiccant-based HVAC applications.

The key elements to the success of the new applications are:
- High electric rates.
- Low cost sources of heat for regeneration.
- High latent loads.

The advantages which desiccant systems have to offer are:
- Reduced operating costs.
- Lower initial system costs for ductwork and fans.

The future of desiccant systems is quite promising and is dependent upon:
- Research for improved efficiencies and low regeneration temperature requirements.
- Utility rate structure changes
- The ability of HVAC manufacturers to produce competitively priced, easily maintained equipment.
- The ability of design engineers to apply desiccant systems in a way which takes advantage of separating the latent and sensible portions of the air conditioning load.

REFERENCES

Schlepp, D., and Schultz, K. *High Performance Solar Desiccant Cooling Systems*, SERI/TR-252-2497, Solar Energy Research Institute, Golden, Colorado, 1984.

Meckler, Gershon. *Techniques for Energy Efficient Integration of Desiccant Dehumidification*, Prepared for Second International Congress on Building Energy Management, May 1983.

American Solar King. Sunair™ Solar Powered Air Conditioning. Leaflet, American Solar King Corporation, Waco, Texas.

Anderson, W.M. *Desiccant-Based Air Conditioning for Commercial Buildings*. Eleventh Energy Technology Conference.

Banks, Paul J. *Utilization of Desiccant Dehumidification for Charles River Barrier Room HVAC Systems*. Paper prepared for Charles River Breeding Laboratories, Inc., Wilmington, Massachusetts.

Cargocaire Engineering Corporation. *The Dehumidification Handbook,* Cargocaire Engineering Corporation, Amesbury, Massachusetts, March, 1982.

Cargocaire Engineering Corporation. Conventional Air Conditioning Versus Superaire®–251. Boston, Massachusetts. Comparison prepared for XENERGY, Inc., April 1984.

Cargocaire Engineering Corporation. *Superaire Cuts Supermarket Cost by Cutting Frost.* Leaflet, Cargocaire Engineering Corporation, Amesbury, Massachusetts, 1984.

Mann, Todd S. "Supermarket Energy Use and Expense," Research Division Food Marketing Institute, Washington, D.C., 1982.

Penney, T.R. and Maclaine-Cross, I. *Advances in Open-Cycle Solid Desiccant Cooling,* Solar Energy Research Institute, Golden, Colorado.

21
Radiant Heat for Affordable Comfort

R.D. Watson

INTRODUCTION

Radiant, or infrared, heating is not new; however, there is a renewed respect for radiant advantages both as primary and backup heating as architects, engineers and builders increasingly incorporate extensive solar design, skylights and glass in their residential, commercial and institutional building.

Unlike conventional, convective systems, radiant heat satisfies the three main factors controlling heat loss from the human body — radiation (electromagnectic waves, or warmth, radiating from the body), convection and evaporation. The major loss, and physiologically the most important, is due to radiation. In spite of this, it rarely receives consideration in any type of air conditioning system and is entirely ignored in all types of convected heat systems.

According to the ASHRAE Systems Handbook, 1976, "We must conclude, therefore, that the usual methods of heating and cooling are basically inadequate, since no system can produce conditions compatible with the physiological demands of the human body, unless radiation losses are satisfied in some way.

"It is sometimes claimed that a radiant heat system is desirable for only certain buildings and in some climates, but not desirable otherwise. This is undoubtedly due to lack of understanding . . ."

BACKGROUND

The invigorating effect of exposure to the sun's rays on a cool, but sunny, day is familiar to everyone. Radiant heat produces a comfortable feeling in much the same way by heating people and objects first, which reradiate that warmth, gradually warming the surrounding air. ASHRAE studies indicate that comfort can be achieved at a 6°-8°F lower ambient air temperature setting than convective systems. There are no chilling drafts with radiant, and ceiling-to-floor temperature varies only 2°-4°.

Just like the sun, radiant heat is clean, noiseless, odorless, requires no useable space, and is almost allergy free, as no foreign particles are introduced into the environment or stirred up by the system.

Radiant heat systems generate operating cost savings of 20%-50% annually compared with convective systems. This is accomplished through room-by-room temperature control and permanent setback due to equivalent comfort at lower thermostat settings. Savings are also due to the superior, cost effective design inherent in direct source-to-object radiant heating products.

The Department of Energy concurs with generally accepted studies that 3%-4% is saved for each degree the thermostat is lowered (see Figure 21-1). Users of surface-mounted radiant panels take advantage of this fact in two ways. First, they are able to achieve comfort at a lower ambient air temperature, normally 60°-64°F as compared with convection heating air temperatures of 68°-72°F. Second, they are able to practice comfortable day and night temperature setback—usually 58°F in areas used frequently, 55°F in areas occasionally used, and 50°F in those seldom used.

COMPARATIVE ANALYSIS

Convective systems are designed to heat the room or entire building. Any heating system which primarily uses the air as the heat transfer medium is a convective heating system. Normally the heat transfer source or outlets are positioned at the perimeter on the outside walls of the room to be heated. Forced air systems usually

Figure 21-1. Percentage reduction in annual heating load
resulting from lower setting of thermostat*

Original Thermostat Setting °F	Degree Decrease in Original Thermostat Setting									
°F	1°	2°	3°	4°	5°	6°	7°	8°	9°	10°
70	3.74	7.41	11.02	14.56	18.03	21.42	24.74	27.99	31.16	34.26
69	3.81	7.56	11.24	14.84	18.37	21.81	25.19	28.49	31.70	34.85
68	3.90	7.72	11.46	15.13	18.71	22.23	25.65	29.00	32.27	35.46
67	3.97	7.87	11.69	15.42	19.07	22.64	26.12	29.52	32.84	26.10
66	4.06	8.04	11.92	15.72	19.44	23.06	26.60	30.07	33.46	36.76
65	4.14	8.19	12.15	16.03	19.80	23.49	27.10	30.64	34.09	37.44
64	4.22	8.36	12.40	16.34	20.19	23.95	27.64	31.24	34.74	38.13
63	4.32	8.54	12.65	16.67	20.60	24.45	28.21	31.86	35.41	38.86
62	4.41	8.71	12.91	17.02	21.04	24.97	28.87	32.49	36.09	39.61
61	4.50	8.90	13.19	17.40	21.51	25.50	29.38	33.15	36.83	40.40
60	4.60	9.11	13.51	17.81	21.99	26.05	30.00	33.85	37.59	41.20
59	4.72	9.34	13.85	18.23	22.48	26.62	30.66	34.58	38.36	
58	4.85	9.58	14.18	18.65	22.99	27.23	31.34	35.30		
57	4.97	9.80	14.50	19.06	23.52	27.84	32.01			
56	5.09	10.03	14.83	19.52	24.06	28.45				
55	5.21	10.27	15.21	19.99	24.62					

*Results assume no internal or external heat gains. This chart is applicable to residences and commercial buildings where these gains are minimal.

Example: Setback from 70°F to 60°F would result in an annual heat load reduction of 34.26%, as shown in the circled numbers.

involve placement of hot air ducts on the outside floor or wall and return cold air ducts on the opposite wall. With this system, cold air is sucked out of a given area and replaced by warm air which rises to the ceiling, warming the wall and the ceiling in the process and gradually losing heat and falling down to the cold air return.

Fin tube central hot water "radiant" systems, are largely convective heaters warming the air which rises along the exterior walls in much the same way as a forced air system. They lose considerable heat to the adjacent exterior wall which may well be exposed to below 0°F temperatures only 6"-8" away. An interesting way to view this conventional heating approach is to consider how much lighting power would be needed if we tried to light rooms by placing all lighting along the outside walls at floor level behind or under furniture and draperies.

Another costly characteristic of hot water or forced air systems is considerable heat transmission loss through ducts or pipes used to convey heat to the area to be warmed. In addition, the heater or boiler must initially be warmed resulting in significant heat loss. For gas systems, efficiency averages 60%-80% and oil systems average 55%-65%, depending upon burner age, adjustment and condition. The newer gas hydro-pulse furnaces have a high combustion efficiency, but this must be offset by the considerable increase in power consumption resulting from the electric arcing process essential to system performance. Maintenance, which is normally low for a gas burner, increases significantly for the high efficiency furnace. In general, the recovery response time of these systems is relatively long and results in use of the entire furnace no matter how small the heating area requirement might be. Therefore, temperature setback is not very meaningful or practical.

Heat pumps, in addition to the normal convective disadvantages, are drafty because the heated air they produce is about 80°-95°F; therefore, a larger volume of air must be circulated to satisfy the heat load. Efficiency declines markedly as the temperature drops below 35°F. Maintenance is a major annual cost factor. The argument that you have both heating and air conditioning is not compelling, as historically, heat pumps have been approximately 15% more expensive in terms of lifecycle costs than stand-alone central air conditioners. For a variety of reasons room-by-room temperature control is neither practical nor cost effective.

PRODUCT ARRAY

Radiant systems vary from oil or gas hot water piping in the floor or ceiling, to electric coils, wiring, gypsum panels with Nichrome wire, flexible elements in the ceiling—either on or above the gypsum board, metal panels generally mounted by means of a bracket about an inch below the ceiling surface and fiberglass heat modules mounted directly to the ceiling surface, several models may be dropped directly into a T-Bar grid. The radiant heating industry has been expanding in recent years. Consumers now have a broad range of products available to choose from.

Each radiant system affords the economics of individual zone control, no maintenance, and 100% efficient use of electricity. For comfort you have to feel to believe and cost you can afford, zoned electric radiant heat deserves a close examination.

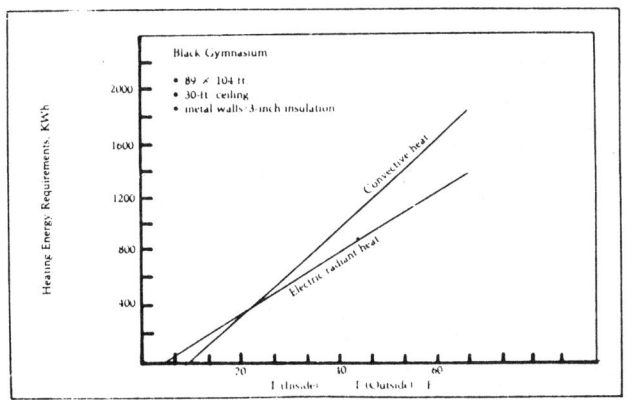

Figure 21-2. Electric infrared system vs. convective heat system.

COMMERCIAL PERIMETER
RADIANT HEATING MAKES SENSE

In the review of commercial building heating and cooling requirements, focus centers on determining design heat loss and occupied cooling load. By briefly walking through a typical case analysis, it is easy to see why Solid State Radiant Heating Modules should be the obvious heating choice.

The first figure traditionally examined is the design heat loss, which is the heat loss under conditions of an empty building with absolutely no contributory heat gains under design (worst case) conditions. Normally, in one way or another, this heat loss must be demonstrated to be met for local building code purposes.

In the real world, heat from people, lighting, computers, etc. routinely supply 50% or more of the heat needed during the most severe weather. When the outdoor temperature is above 30°F (virtually 70% of the time even in northern climates), heating from people, lights, and power, supplies too much heat. As the following analysis illustrates, cooling a commercial building is the problem. Commercial buildings heat themselves. Supplementary heating is all that is required. The unfortunate fact is that most commercial buildings have grossly oversized heating plants that waste enormous amounts of energy and money.

Description of Assumptions

The case analysis we are examining is the top floor of a commercial office building. The floor area is 10,000ft^2. The top floor zone was chosen because for this particular building it has the highest heat loss rate per square foot of occupied floor space. In the winter we are maintaining 70°F indoor temperature at 0°F outdoor temperature, 15 mph wind. In the summer we are maintaining 78°F indoor temperature at 95°F dry bulb, 77°F wet bulb, 7½ mph wind. The building is located in Westport, Connecticut, which has approximately 5,800 annual degree days. We are assuming 100 people will be occupying the space and that lighting and power heat gains will be about 3 watts per square foot. Fresh air is being introduced mechanically at approximately 7½ CFM per person.

Basic Analysis

We constructed a heating/cooling load profile (Figure 21-3) for the 10,000ft^2 space, which relates graphically the various heating and cooling demands for every temperature between 0°F and 95°F for both occupied and unoccupied cycles. Line S-R is the Occupied Heating Line, Line P-G is the Unoccupied Heating Line, Line T-L-E is the Cooling Line. Heating BTUH's are read on the vertical axis below temperature scale.

The load profile can be used to analyze many things. However, for this discussion we are focusing on heating the building at 0°F. The broken line B-C represents the heating load requirement considering the skin losses and fresh air requirements, but no credits are

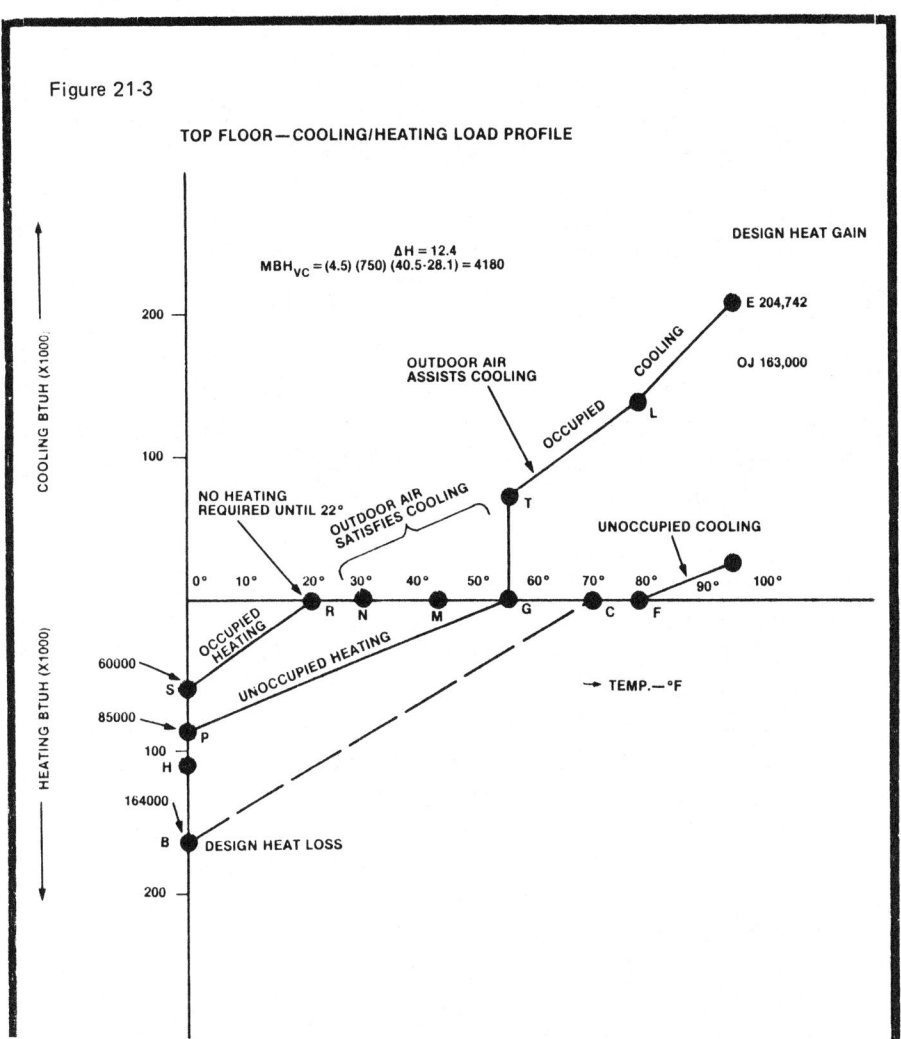

Figure 21-3

taken for people or lighting. It intersects the heating BTUH axis at point B (164,000 BTUH). This value will be the number that is typically used to size the heating equipment for building permit purposes. It is the number most heat loss calculation procedures will produce. It works out to be 16.9 BTUH FT^2 or Watts FT^2, typical for today's well insulated building. Moving up the heating axis, the next important value we come to is point P, the unoccupied heating requirement of 85,000 BTUH on a 0°F day. Here we are allowing the building temperature to float down to 55°F (point G); the mechanical fresh air systems are assumed to be turned off. This load is slightly less than 50% of the calculated design heat loss value.

Finally, moving further up the heating axis, we come to point S the occupied heating requirement at 0°F, including the fresh air load, but also allowing for heating contributions from people, lights, and power. It's value of 60,000 BTUH, a little more than 1/3 the value usually used to size heating equipment. Surprised? Follow along the Occupied Heating Line until you come to Point R. Read 22°F on the Temperature Scale. Point R represents the temperature where this building no longer needs air conditioning, but begins (emphasis on begins) to need heating. This building actually requires air conditioning down to 22°F, when it is fully occupied!

Conclusion

When working through the above analysis one must assume a heating and air conditioning system that can take advantage of the internal heat loads.

Heating modules located on the perimeter of the building can be part of a system designed to closely approximate the ideal situation. A central air conditioning system, typically a variable air volume system, handles the cooling for the interior zones and supplies tempered fresh air to the perimeter at all times. The radiant modules supply heat only where it is needed, and when it is needed. There can be as many thermostats as there are people or offices, if so desired. Operating costs will be low, because as Figure 21-1 reveals, very little heat is required to begin with. No monster boilers maintaining steam or water temperature occupying floor space sending dollars up the chimney; no oversized gas-fired roof tops radiating heat to everything but the inside of the building.

Radiant modules represent an optimum approach to heating commercial office space: low first cost, efficient use of energy, flexible zoning and easy adaption to changing office arrangement and floor plans.

ZONE CONTROL

Effective Zone Control

Zone control is the decentralization of the comfort system into defined areas, independently supplied and controlled by individual control units. (See Figure 21-4.) Two fundamental components are required for a zoned system to be effective. First each zone must have a device that monitors, controls and maintains the desired comfort level of the zone. Second, the system must respond rapidly to changes in demand to prevent discomfort while change from one level to another takes place.

The solid-state heating system effectively meets these requirements. First, it is decentralized, with each zone having its own thermostat which controls and monitors a module or group of modules to maintain the desired comfort level within the zone. Second the desired comfort level is reached within minutes because each module responds as soon as it is energized. The occupant is immediately warmed, independent of the surrounding air temperature.

Detailed testing provided that the same temperature level is reached three times faster with radiant heat than electric baseboard heaters.[1]

When mounted to the ceiling, the radiant heat module is thermally insulated from the ceiling mass to give the desired quick radiant response with minimal back and side losses when compared to other radiant systems. These factors combine to provide a rapid sense of comfort from the radiant heat module with 4-5 minutes from the time it is energized. This is supported in the 1979 report by the National Bureau of Standards to the Department of Energy which called "well suited to zone heating" and recommended the system for funding under the Energy Related Inventions Program.

Auxillary Heating With Enerjoy Adds Comfort To Savings

By combining lower overall ambient temperatures with Enerjoy placed in frequently used rooms or areas (for example, the bathroom,

---

[1] "Baseboard Heaters and Ceiling Mounted Panels: A Comparative Evaluation, Environmental Research Institute, Kansas State University, 1979."

kitchen and family room), comfort can be achieved in those rooms while significantly reduced temperature settings can be maintained in areas which are infrequently occupied. The energy load savings potential of 3 percent per degree of setback will result in sizeable savings with occupants much more comfortable as well.

Zone Control Offers Greatest Savings Potential In Commercial Applications

Office buildings and other commercial structures, especially those with mixed tenancies and usages are prime candidates for zoned heating. Total zone control provides maximum flexibility to respond to varying solar loads, wind chill factors and wide variations of building usage on a room-by-room basis. Commercial or retail spaces combined with warehouse or storage facilities should always be zoned, for storage areas need only be maintained at temperature levels necessary to prevent inventory damage without affecting normal comfort conditions of the sales area. (See Figure 1.)

Multi-room facilities such as hotels, motels and schools face the problem of intermittent use and varying degrees of solar loading as the sun changes position throughout the day. With a zoned system, each room is provided with the capacity to maintain the highest level of comfort at the lowest cost as these changes take place.

Multi-family dwelling units also greatly benefit from fully zoned systems. Each tenant-owner retains control of his energy usage and decides for himself how his energy dollars should be spent to best suit his customary life style. Rental units can be separately metered to equitably distribute utility costs to each tenant, thereby limiting the energy cost liability of the owner.

Well Designed Control System Assures Zone Control Efficiency

The properly designed and installed control system plays an important role in any zoned comfort system. In addition to the availability of good line and low voltage thermostats, new technology now makes it possible for comfort zones to be automatically advanced to warm-up before use, or set back to lower levels after use with solid-state "smart" thermostats which can be programmed according to need. School classrooms and office buildings are partic-

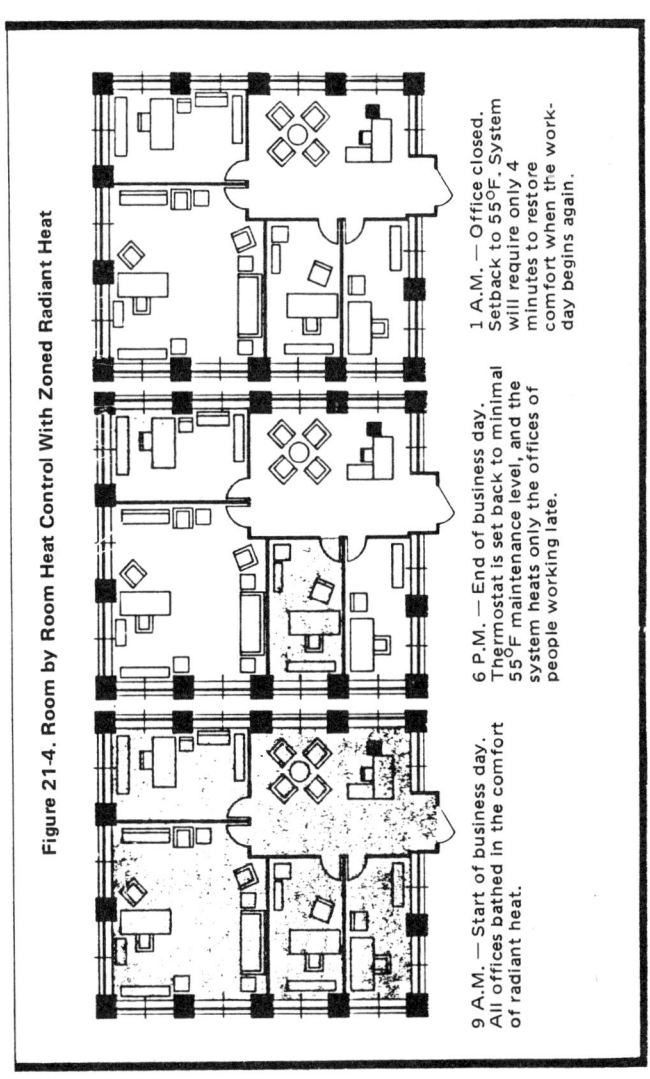

Figure 21-4. Room by Room Heat Control With Zoned Radiant Heat

9 A.M. — Start of business day. All offices bathed in the comfort of radiant heat.

6 P.M. — End of business day. Thermostat is set back to minimal 55°F maintenance level, and the system heats only the offices of people working late.

1 A.M. — Office closed. Setback to 55°F. System will require only 4 minutes to restore comfort when the workday begins again.

ularly good applications because of predictable routines. Also available are individual room thermostats which can be controlled from a central monitor.

Proper Design of System and Sizing Layout
Assures Efficient Zone Control

Designing the solid-state heating system for efficient zone control begins with a heat loss analysis calculated room by room and includes degree days for the particular geographical location, number and location of doors and windows, how much and what type of insulation and other factors. Each room must be sized and treated as an individual unit for correct design.

Compared to a central system, where all interior walls are kept at the same temperature, some heat transfer may take place from room to room in a zoned system where an occupied warm room is adjacent to an unoccupied colder one. Also a chilled room surrounded by other cold rooms will require a slightly greater capacity to recover and maintain a desired comfort level than will a room adjacent to other heated areas. This heat transfer, depending on room location, can vary as much as 10-15 percent and should be considered when sizing a room to meet minimum comfort levels under 'worst case' conditions. When it is properly sized, optimum performance and most desirable comfort at greatest operating economy may be achieved with Enerjoy.

In conclusion, the quick 4-minute comfort response of radiant heat modules offers the design engineer the opportunity to provide exactly the heat required so that occupants pay only for *comfort* heat when, where, and as needed in commercial, institutional, and residential buildings.

22

Spreadsheet Models to Determine HVAC System Savings

D.L. Pope

ABSTRACT

Energy savings calculations which use only one or two load points at average or full load conditions can give misleading results, especially if more than one conservation measure is being evaluated. A more accurate method is presented using a computerized spreadsheet model of building part load conditions as a function of standard hourly bin weather data. The full range of building operation from heating through cooling conditions are examined. Individual loads, equipment and their operating characteristics are tabulated against changing load conditions and can be adjusted individually to reflect changes caused by energy conservation measures. In addition, the tabulated load components may be adjusted to reflect the interactive effects of system changes. The full range of building heating and cooling loads, and equipment part load operations are then considered for a more accurate calculation of savings. This technique is based on ASHRAE's "Simplified Energy Analysis Using The Modified Bin Method."

Degree Day Method

The traditional method for estimating energy use has been the Degree Day Method. In 1932, the American Gas Association and National District Heating Association developed the Degree Day Method based on 65 degrees for estimating annual fuel bills on residential buildings. This is referred to as a "Single Measure" method since only one calculation is made. The temperature where the addition of heat is required in residential buildings, known as the balance point, was determined to be close to a constant 65°F. This reflected building techniques of the day which included poor insulation, relatively low internal heat generation, loose fit windows, and construction with high infiltration rates. Fuel consumption was estimated by taking the difference between 65° and the 24-hour average of the day's temperature and accumulating these degree days into a one-year total. This total was multiplied by the peak building load/hour, 24 hours and various correction factors, then divided by the design temperature difference, fuel efficiency, and BTU/fuel unit. The resulting answer was claimed to be within 20% of actual fuel bills.

A similar method uses Cooling Degree Days to estimate annual required cooling energy in residential buildings. A cooling degree day is an average daily temperature one degree above 65 degrees. Figure 22-1 illustrates the basic relationship between outside temperature and degree day calculated load. Note that the load line is a wide line covering a typical daily range of plus or minus 10 degrees with balance points as low as 45 degrees.

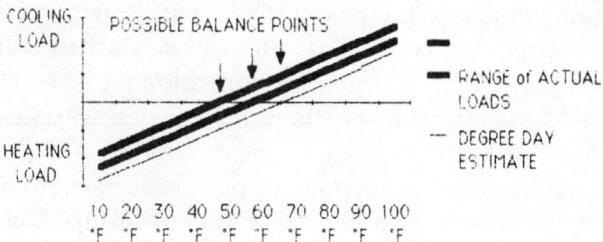

Figure 22-1. Comparison of the degree day model vs. range of actual loads.

A number of inaccuracies can occur, due primarily to assumed "average" values. First, the Degree Day method was intended for use in residential buildings with internal loads typical of 1930. Today's residential internal electric loads are typically 15 times the level of the 1930's. Second, this method will not address internal loads and occupancy characteristics of larger commercial buildings. A modern building's balance point can vary considerably depending on internal load and the amount of outside air introduced. It is possible for this balance point to be in the range of 35 to 65 degrees Fahrenheit. Since only the average daily temperature is used, substantial error can result on days when the daily average is 65 or above, but a nighttime low temperature swing can require heating. The farther below 3,000 degree days for a given region, the more error will occur due to partial heating days. It is easily possible to over-estimate fuel consumption by 100% or under-estimate by 50%. Fuel burning equipment efficiencies decrease considerably at partial load. An 80 percent efficient furnace typically has an overall average efficiency of 55%, but this can vary from 33% to 77% for a 65% confidence range (one standard deviation).[1] Attempts have been made to develop correction factors, but the wide range of error possible in these factors limit their usefulness.[2] Occupancy patterns for commercial buildings also contribute significantly to fuel estimation error. The Degree Day Method assumes 24-hour occupancy. Offices are not occupied during colder night hours and will not use as much energy as a residential occupancy would. The Heating Degree Day Method is totally unsatisfactory as an energy estimation method for large commercial buildings.

Using the Cooling Degree Day Method as an energy estimation tool has all of the above problems plus some. In the Cooling Degree Day Method, the cooling load is assumed to be dependent only on average daily outside temperature. Daily extremes are not considered. Neither are solar gains through walls and windows, nor latent cooling loads from outside air and people. Latent load alone can exceed sensible loading in a 100% outside air system. Cooling Degree Days is an even less satisfactory energy estimation method than Heating Degree Days and again is totally unsatisfactory as an estimation Method for a large commercial building.

Bin Method

Most of the problems with Degree Day energy estimation methods have been the results of trying to estimate a wide range of operating conditions with a single "average" calculation. Much of the error caused by averaging part load operation can be reduced by multiple load calculations. The bin method uses multiple calculations across the range of outside hourly temperature bins, in association with the corresponding equipment part load characteristics. For a number of years, a simple version of the bin method has been used to calculate the average annual performance of electric heat pumps. Since a heat pump's coefficient of performance changes dramatically with outside air temperature, this method was necessary to get a realistic estimate of power consumption over a year.

Standard hourly bin data, for many locations, is available from Air Force "Manual 88-29, Engineering Weather Data." Copies can be obtained from the Government Printing Office. This weather data is normally grouped into total number of hours per bin temperature range by month with an annual total. Each day is divided into 3 time groups: Midnight to 8 a.m.; 8 a.m. to 4 p.m.; and 4 p.m. to midnight. The second group and a fraction of the third group may be taken as weather data for the occupied period of an office. Figure 22-2 below is a histogram of weather data showing total hours and office occupancy hours from 8 a.m. to 6 p.m. in the Atlanta area.

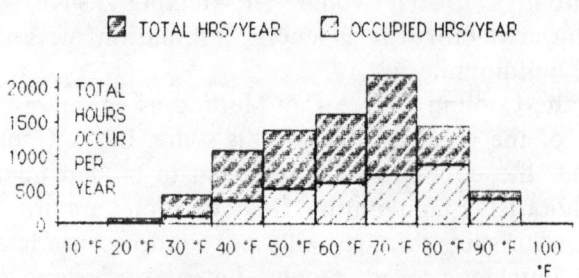

Figure 22-2. Histogram of average weather data for Atlanta, Ga.

Modified Bin Method

In 1983 ASHRAE published "Simplified Energy Analysis Using the Modified Bin Method" which presents a detailed methodology of estimating energy use relative to weather data. A building's heating and cooling load is broken into its component loads that are calculated for their average diversified values rather than peak load conditions. Generally two sets of calculations are performed, one for occupied and one for unoccupied periods. When completed, this load profile will resemble that of Figure 22-3. These loads are then taken as a percentage of full load boiler or chiller capacity. Figure 22-4 and Figure 22-5 show the relationships between part load operation and energy consumption for efficient boilers and chillers respectively.

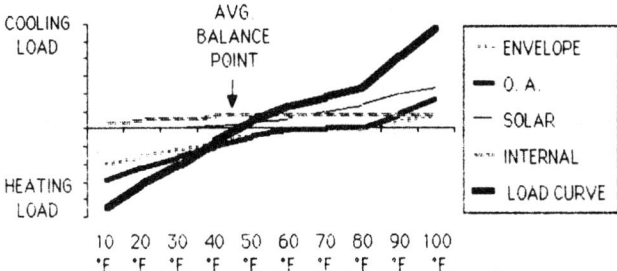

Figure 22-3. Modified bin model showing individual load components and the combined load curve.

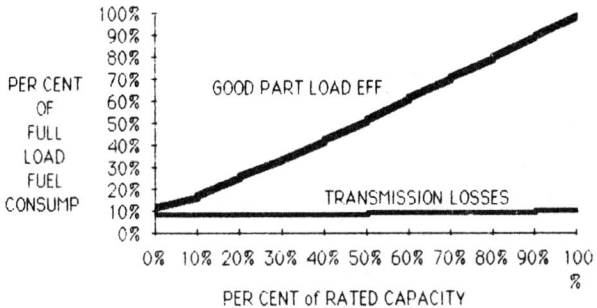

Figure 22-4. Boiler part load fuel consumption.

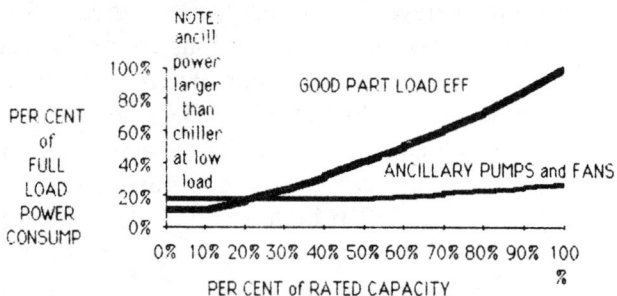

Figure 22-5. Chiller part load power consumption.

A separate calculation is performed at each bin temperature. The total load is determined for each bin, its percentage of full load capacity of cooling or heating and its resulting electric or fuel consumption. The hourly fuel consumption multiplied by the hours per year yields the annual consumption in each bin. Totaling all bin calculations determines total annual consumption. If a significant load exists during unoccupied hours a second set of calculations must be performed for the unoccupied period. Both the occupied and the unoccupied calculations are totaled to determine annual consumption. If monthly bills are to be calculated to determine actual cost on a seasonal utility rate, the calculations must be repeated 12 times using monthly bin data.

SPREAD SHEETS

Personal computers offer a powerful tool for solving such tedious calculations as discussed above: spread sheet programs. ASHRAE's "Simplified Energy Analysis Using the Modified Bin Method" proposes that the engineer write a specific computer program for each building model. Few practicing engineers have the time, budget, experience or desire to write and de-bug such programs. Most, however, are capable of entering values and formulae on a spread sheet.

Load curves represented in Figure 22-3 are defined as a function of the bin temperature midpoints. These functions are linear or

Spreadsheet Models to Determine HVAC System Savings   383

polynomial equations. These equations may be generated with a curve fitting program from actual calculated data. Likewise, the hourly fuel consumption of boilers or chillers are tunctions of the percent of full load capacity. Again these functions are defined by equations for the curves shown in Figures 22-4 and 22-5. These curves are fitted to actual test data from the specific equipment being studied. Once the equations for one bin line have been defined they are simply copied down to all other bin lines. The totals are obtained by defining the cell at the bottom as the sum of the column of numbers above it. Once this has been done you have "template" models of the building being analyzed. They will appear as follows in Table 22-1A and Table 22-1B.

	B	C	D	E	F	G	H	I
1				TABLE 1.A				
2	MODIFIED BIN SPREADSHEET ANALYSIS OF A BUILDING IN ATLANTA, GA. - SUMMER							
3	BIN	MEAN	FREQ.	AVG TONS	% TOTAL	ANCILL.	CHILLER	TOTAL
4	TEMP	COINC.	OCCUR.	DIVERS.	COOLING	COOLING	KW PER	COOLING
5	RANGE	W. B.	HRS/YR	COOLING	CAPACITY	LOAD	BIN	KWH PER
6	(degrees F)		occupied	LOAD	(300tons)	KW		BIN
7	55 - 60	52	773	20	7%	50	39	68,797
8	60 - 65	57	845	40	13%	50	50	84,500
9	65 - 70	62	986	76	25%	51	77	126,208
10	70 - 75	67	1,201	111	37%	51	104	186,155
11	75 - 80	69	839	147	49%	71	151	186,258
12	80 - 85	70	612	183	61%	72	179	153,612
13	85 - 90	72	367	219	73%	72	207	102,393
14	90 - 95	74	135	254	85%	72	233	41,175
15	95-100	74	20	290	97%	74	262	6,720
16	TOTALS		5,778		TOTAL ANNUAL KWH			955,818

Table 22-1A.

	B	C	D	E	F	G	H	I
1				TABLE 1.B				
2	MODIFIED BIN SPREADSHEET ANALYSIS OF A BUILDING IN ATLANTA, GA - WINTER							
3	BIN	MEAN	FREQ	AVG mbh	% TOTAL	HEATING	TOTAL	TOTAL
4	TEMP	COINC	OCCUR	DIVERS.	HEATING	TRANSM.	HEATING	HEATING
5	RANGE	W B	HRS/YR	HEATING	CAPACITY	LOSS	mbh load	MCF GAS
6	(degrees F)		occupied	LOAD	(2,000mbh)	(mbh)	per BIN	per BIN
7	5 - 10	6	1	1,800	90%	120	1,920	2
8	10 -15	11	9	1,636	82%	118	1,754	20
9	15 - 20	15	23	1,473	74%	116	1,589	50
10	20 - 25	20	51	1,342	67%	115	1,457	106
11	25 - 30	24	134	1,145	57%	113	1,258	260
12	30 - 35	29	303	982	49%	111	1,093	540
13	35 - 40	34	471	818	41%	109	927	749
14	40 - 45	38	608	655	33%	107	762	830
15	45 - 50	42	665	491	25%	105	596	736
16	50 - 55	47	709	327	16%	104	431	582
17	55 - 60	52	773	164	8%	102	265	397
18	60 - 65	57	845	0	0%	100	100	164
19	ANNUAL TOTAL		4,592		ANNUAL TOTAL GAS CONSUMP			4,436
20					AVG. ANNUAL SYSTEM EFFICIENCY			45%

Table 22-1B.

These templates may be copied, renamed, and modified to calculate energy use in other similar buildings, or to look at the same building with different equipment or a modified system. The beauty of a spreadsheet is that once the template structure has been created it may be modified with relative ease to play "what-if" for as many alternatives as you wish. Furthermore, intermediate answers may be checked by calculating in stages, as in column "F" in Table 22-1A & 22-1B show how loaded the boiler and chiller are at a given Bin range. When your calculations are complete the spreadsheets depicting various alternatives may be printed out for a neat detailed record of your calculations. Many spreadsheets allow you to print the cell formulas as well.

Methodology

Depending on whether you are modeling a building under design or an existing building you will take one of two approaches to setting up the spreadsheet load curve. If the building is under design or very new the average diversified load model must be calculated. This is done in the manner described above, by developing separate equations for each load component and totaling, or by manually calculating average component loads at three or more points along the Bin range. A more detailed description of this process is given in the ASHRAE book "Simplified Energy Analysis Using The Modified Bin Method."[3] However, if the building is existing, has been operating for 2 or more years, and has good records on energy demand and consumption, a simpler technique should be used. Gas meter consumption data is coordinated with recorded weather data from the National Oceanic and Atmospheric Administration to develop a "Scatter Plot" of consumption versus average outside temperature. An average gas trend line is constructed from this data and used as a load profile. The same method should be used with electrical demand data to produce an electrical trend line. Figure 22-6 on the next page illustrates such a plot for gas consumption. This method of using actual meter history to derive average load curves is preferred when possible since a large amount of guess-work is eliminated.

The curves representing equipment part load performance are best obtained from the equipment manufacturer. A reasonably

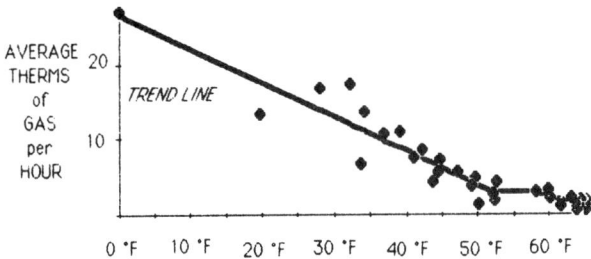

Figure 22-6. Scatter plot of gas concumption vs. average daily weather data.

accurate curve can be constructed from four or five points of part load operation. Some curves may prove difficult to fit to an equation, or you may not want to go through this step for a simple analysis. In such a case, the data points of the spreadsheets are mannually determined by interpolation from a manufacturer's chart. The manually derived points are then entered into the spreadsheet without generating equations from the performance curves. By example, in Table 22-1A above, the cooling kW consumption column could be manually calculated using the manufacturer's performance charts, at the percent load shown instead of fitting an equation to the manufacturer's chart. This is one of the chief advantages of a spreadsheet: you can observe the calculations in stages and manually perform part of them. This is also of great advantage when a system modification has interactive effects. For example, if you are simulating a water side economizer, the chiller kW savings should be offset to some degree by the extra kWh consumption of the cooling tower fan and the condenser water pumps, especially if higher pumping head results from heat exchangers or filters. Each of these pieces of equipment should be listed in a separate column, for clarity.

When examining energy conservation alternatives, a base case of the building must be used for comparison so that a net energy and cost savings may be determined. When utility data is available, it should be used to check the accuracy of the spreadsheet model. The actual fuel consumption should be within 10 to 15% of the total value calculated by the spreadsheet. If several years' data is available, the average should be within 5% to 10%, since weather effects have been reduced.

Applications

When applying this technique, it is implicit that the more itemized and detailed the spreadsheet, the more accurate the results and conclusions. For example, if the building load curve were broken into its individual components as represented in Figure 22-3, the solar load component may be reduced by the amount reflected by window film. The spreadsheet bin model would show not only the cooling energy saved but also the extra heating energy required due to less winter solar gain. If the solar component were broken down even further into north, east, west and south walls and windows, more specific solar energy conservation opportunities may be examined. For example, solar film on a south wall may not be practical. The internal load component should be adjusted to reflect the effects of daily occupancy as well as low occupancy on weekends.

Some of the most significant savings have been found by modifying the primary boiler and chiller systems and their secondary distribution systems. This calculation method reveals how significant part load operating efficiencies are with respect to total annual energy consumption. Most of the time during the year, heating and cooling equipment is operating at a low part load range. Figure 22-7 shows the total energy consumed per year by bin range. Note that most energy is consumed in the moderate temperature ranges where boilers and chillers are loaded at less than 50% capacity. This is more a result of the higher number of hours in these bin ranges than load or equipment efficiencies.

Primary boilers and chillers with good efficiency at part load conditions and a good turndown ratio can provide significant savings and payback. Multiple boilers and chillers improve this further by allowing even better turndown ratios. Water and air distribution systems that utilize variable volume operation at part load provide some of the best savings in these temperature bins. Figures 22-8 and 22-9 graphically demonstrate these relationships.

Somewhat more involved are the techniques used for modeling various control concepts of air distribution systems and chilled water/

Spreadsheet Models to Determine HVAC System Savings 387

hot water reset. The referenced book "Simplified Energy Analysis Using The Modified Bin Method"[3] goes into these techniques with great detail.

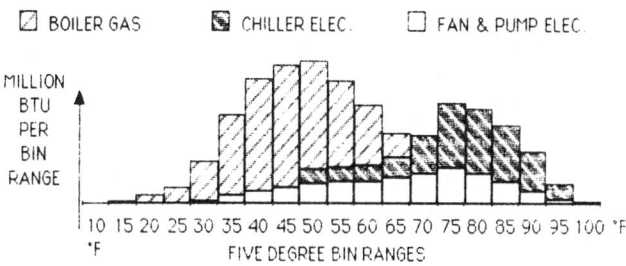

Figure 22-7. Example of BTU's consumed per bin range in an Atlanta, Ga., building.

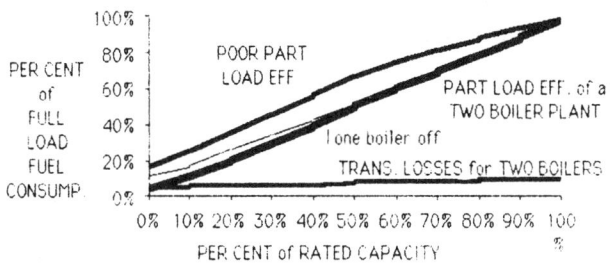

Figure 22-8. Part load fuel consump. for different boilers and configurations.

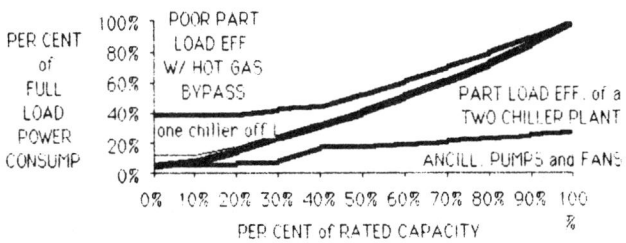

Figure 22-9. Part load power consump. for different chillers and configurations.

SUMMARY

Using computerized spreadsheets to perform a Modified Bin Analysis on a building is an effective and accurate tool for estimating energy consumption of commercial and institutional buildings. It is not limited to small residential buildings, as is the Degree Day Method. Instead, it is as flexible and complex as the analysis requires. If properly set up it can produce answers approaching the accuracy of large main frame computer programs that calculate hourly values for the full year. As with any analysis the complexity must be kept to reasonable limits given the expected accuracy of estimated energy consumption. Unpredictable parameters, such as changing occupancy patterns and weather variations, can cause even the most complex main frame program analysis to vary from actual recorded usage. It must be understood that the value of a building model is not necessarily its accuracy in predicting utility bills. Its greatest value is in predicting energy savings from various proposed alternatives. Modified Bin analysis using computerized spreadsheets is an excellent tool for this use.

REFERENCES

[1] ASHRAE, "1985 Fundamentals," Pg. 28.9

[2] ASHRAE, "1976 Systems," Pg. 43.8

[3] ASHRAE, Prepared by David E. Knebel, "Simplified Energy Analysis Using The Modified Bin Method," 1983.

Index

A
Absorption Chillers 317
Affinity Laws for Pumps 27
Air Changes Per Hour 3
Air Change Method 10
Air Source Heat Pumps 98
ASHRAE Standard 62-1981 303

B
Bin Method 379
Boiler Controls 243
Building Dynamics 4

C
Case Study: Heat Pump Heat Recovery System at Tend-R-Fresh 85
Case Study: Heat Pump Strategy at the Nevada Test Site 97
Case Study: Thermal Energy Storage for Municipal Buildings 167
Case Study: Three Proposed Cold Thermal Storage Systems for a U.S. Navy Shore Facility 143
Centrifugal Chillers 317
Chilled Water Storage 108, 129
Chiller Controls 242
Closed Vapor Compression Cycle 45, 46
Computer Modeling 131
Conductance 2
Constant Volume System 33
Cooling Tower Optimization 273
Cost Comparison of HVAC Controls 251
Counter Flow 283, 288
Crack Method 11
Cross Flow 283, 288

D
Degree Day Method 377
Degree Days 1
Desiccant Based Air Conditioning Systems 349
Direct Digital Controls 241, 250
Direct Evaporative Coolers 191
Distributed Intelligence 261
Distributed Processors 262
Dry Bulb Economizer 36
Dry Bulb Start Time Optimization 343
Dry Towers 282, 289
Dual Duct System 30, 34

E
Earth Coupled Heat Pump 69
Economizer Cycle 37
Electric Utility Schedules 123
Enthalpy Control 38
Enthalpy Start Time Optimization 343
Eutectic Salt Storage 129
Evaporative Cooling 191
Evaporative Roof Spray Cooling 217
Evaporator Technology 61

F
Fan Coil Unit 31
Fan Distribution Systems 20
Fan Laws 22
Fan Performance Curves 23
Fluid Flow 26

H
Heating Capacity of Air 4
Heat Pump Applications 43
Hot Water Converters 32
Humidification Steam 333
HVAC Controls 241, 243

I
Ice Storage System 108, 129

Indirect Evaporative Cooling 191, 192
Induction System 31, 33
Interface Controls 245

L
Latent Heat 9, 219

M
Modified Bin Method 380
Multizone System 30, 36

O
Open Cycle Heat Pumps 57

P
Perimeter Radiation 32
Plate Type Indirect Coolers 206
Pneumatic Controls 249
Psychrometric Chart 14, 346

R
Radiant Heat 365
Radiant Heat Gain 220
Resistance to Heat Flow 2
Rotary Heat-Exchange Coolers 204

S
Salt Storage 109
Sensible Heat 9
Single Zone Systems 29, 35
Sizing Thermal Storage System 110

Spread Sheet Models 377
Start Time Optimization 343
Storage Media 125
Sunlight & Glass Considerations 13

T
Terminal Reheat System 30, 37
Thermal Storage
   Benefits 121
   Options 123
Tubular Heat-Exchange System 200

U
Unit Heater 32
Unit Ventilator 31

V
Vapor Compression Evaporation 62
Variable Air Volume System 31, 33
Volume of Air 3

W
Waste Heat Driven Rankine Cycle 45
Waste Heat Sources 48
Water Source Heat Pumps 99
Water Spray Atomization 335
Wet Cooling Towers 273

Z
Zone Controls 244